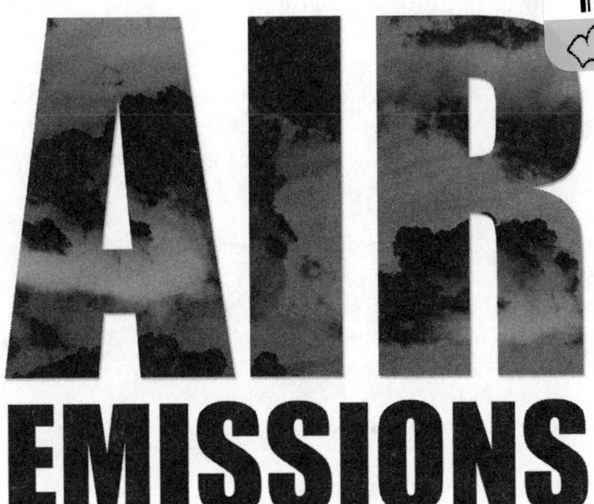

EMISSIONS

From Animal Feeding Operations

Current Knowledge, Future Needs

Ad Hoc Committee on Air Emissions from Animal Feeding Operations
Committee on Animal Nutrition
Board on Agriculture and Natural Resources
Board on Environmental Studies and Toxicology
Division on Earth and Life Studies

NATIONAL RESEARCH COUNCIL
OF THE NATIONAL ACADEMIES

THE NATIONAL ACADEMIES PRESS
Washington, D.C.
www.nap.edu

THE NATIONAL ACADEMIES PRESS 500 Fifth Street, N.W. Washington, DC 20001

NOTICE: The project that is the subject of this report was approved by the Governing Board of the National Research Council, whose members are drawn from the councils of the National Academy of Sciences, the National Academy of Engineering, and the Institute of Medicine. The members of the committee responsible for the report were chosen for their special competences and with regard for appropriate balance.

This study was supported by Contract No. 68-D-01-69 between the National Academy of Sciences and the U.S. Environmental Protection Agency and Grant No. 59-0790-2-106 between the National Academy of Sciences and the U.S. Department of Agriculture. Any opinions, findings, conclusions, or recommendations expressed in this publication are those of the author(s) and do not necessarily reflect the views of the organizations or agencies that provided support for the project.

International Standard Book Number: 0-309-08705-8

Library of Congress Control Number: 200310401

Additional copies of this report are available from the National Academies Press, 500 Fifth Street, N.W., Lockbox 285, Washington, DC 20055; (800) 624-6242 or (202) 334-3313 (in the Washington metropolitan area); Internet, http://www.nap.edu

Copyright 2003 by the National Academy of Sciences. All rights reserved.

Printed in the United States of America

// THE NATIONAL ACADEMIES
Advisers to the Nation on Science, Engineering, and Medicine

The **National Academy of Sciences** is a private, nonprofit, self-perpetuating society of distinguished scholars engaged in scientific and engineering research, dedicated to the furtherance of science and technology and to their use for the general welfare. Upon the authority of the charter granted to it by the Congress in 1863, the Academy has a mandate that requires it to advise the federal government on scientific and technical matters. Dr. Bruce M. Alberts is president of the National Academy of Sciences.

The **National Academy of Engineering** was established in 1964, under the charter of the National Academy of Sciences, as a parallel organization of outstanding engineers. It is autonomous in its administration and in the selection of its members, sharing with the National Academy of Sciences the responsibility for advising the federal government. The National Academy of Engineering also sponsors engineering programs aimed at meeting national needs, encourages education and research, and recognizes the superior achievements of engineers. Dr. Wm. A. Wulf is president of the National Academy of Engineering.

The **Institute of Medicine** was established in 1970 by the National Academy of Sciences to secure the services of eminent members of appropriate professions in the examination of policy matters pertaining to the health of the public. The Institute acts under the responsib-ility given to the National Academy of Sciences by its congressional charter to be an adviser to the federal government and, upon its own initiative, to identify issues of medical care, research, and education. Dr. Harvey V. Fineberg is president of the Institute of Medicine.

The **National Research Council** was organized by the National Academy of Sciences in 1916 to associate the broad community of science and technology with the Academy's purposes of furthering knowledge and advising the federal government. Functioning in accordance with general policies determined by the Academy, the Council has become the principal operating agency of both the National Academy of Sciences and the National Academy of Engineering in providing services to the government, the public, and the scientific and engineering communities. The Council is administered jointly by both Academies and the Institute of Medicine. Dr. Bruce M. Alberts and Dr. Wm. A. Wulf are chair and vice chair, respectively, of the National Research Council.

www.national-academies.org

AD HOC COMMITTEE ON AIR EMISSIONS FROM ANIMAL FEEDING OPERATIONS

PERRY R. HAGENSTEIN (*Chair*), Institute for Forest Analysis, Planning, and Policy, Wayland, Massachusetts
ROBERT G. FLOCCHINI (*Vice Chair*), University of California, Davis, California
JOHN C. BAILAR III, University of Chicago, Chicago, Illinois
CANDIS CLAIBORN, Washington State University, Pullman, Washington
RUSSELL R. DICKERSON, University of Maryland, College Park, Maryland
JAMES N. GALLOWAY, University of Virginia, Charlottesville, Virginia
MARGARET ROSSO GROSSMAN, University of Illinois at Urbana-Champaign, Urbana, Illinois
PRASAD KASIBHATLA, Duke University, Durham, North Carolina
RICHARD A. KOHN, University of Maryland, College Park, Maryland
MICHAEL P. LACY, University of Georgia, Athens, Georgia
CALVIN B. PARNELL, JR., Texas A&M University, College Station, Texas
ROBBI H. PRITCHARD, South Dakota State University, Brookings, South Dakota
WAYNE P. ROBARGE, North Carolina State University, Raleigh, North Carolina
DANIEL A. WUBAH, James Madison University, Harrisonburg, Virginia
KELLY D. ZERING, North Carolina State University, Raleigh, North Carolina
RUIHONG ZHANG, University of California, Davis, California

Consultant

MICHAEL OPPENHEIMER, Princeton University, Princeton, New Jersey

Staff

JAMIE JONKER, Study Director
CHAD TOLMAN, Program Officer
TANJA PILZAK, Research Assistant
JOE ESPARZA, Project Assistant
STEPHANIE PADGHAM, Project Assistant
BRYAN SHIPLEY, Project Assistant
PETER RODGERS, Intern
FLORENCE POILLON, Contract Editor

COMMITTEE ON ANIMAL NUTRITION

Gary L. Cromwell (*Chair*), University of Kentucky, Lexington, Kentucky
C. Roselina Angel, University of Maryland, College Park, Maryland
Jesse P. Goff, United States Department of Agriculture/Agricultural Research Service, Ames, Iowa
Ronald W. Hardy, University of Idaho, Hagerman, Idaho
Kristen A. Johnson, Washington State University, Pullman, Washington
Brian W. McBride, University of Guelph, Guelph, Ontario, Canada
Keith E. Rinehart, Perdue Farms Incorporated, Salisbury, Maryland
L. Lee Southern, Louisiana State University, Baton Rouge, Louisiana
Donald R. Topliff, West Texas A&M University, Canyon, Texas

Staff

Jamie Jonker, Program Officer
Joe Esparza, Project Assistant

BOARD ON AGRICULTURE AND NATURAL RESOURCES

Harley W. Moon (*Chair*), Iowa State University, Ames, Iowa
Sandra Bartholmey, Quaker Oats Company, Barrington, Illinois
Deborah Blum, University of Wisconsin, Madison, Wisconsin
Robert B. Fridley, University of California, Davis, California
Barbara Glenn, Federation of Animal Science Societies, Bethesda, Maryland
Linda Golodner, National Consumers League, Washington, D.C.
W.R. (Reg) Gomes, University of California, Oakland, California
Perry R. Hagenstein, Institute for Forest Analysis, Planning, and Policy, Wayland, Massachusetts
Calestous Juma, Harvard University, Cambridge, Massachusetts
Janet C. King, University of California, Davis, California
Whitney MacMillan, Cargill, Incorporated, Minneapolis, Minnesota
Pamela A. Matson, Stanford University, Stanford, California
Terry Medley, DuPont Biosolutions Enterprise, Wilmington, Delaware
Alice Pell, Cornell University, Ithaca, New York
Sharron S. Quisenberry, Montana State University, Bozeman, Montana
Nancy J. Rachman, Novigen Sciences, Incorporated, Washington, D.C.
Sonya Salamon, University of Illinois, Urbana-Champaign, Urbana, Illinois
G. Edward Schuh, University of Minnesota, Minneapolis, Minnesota
Brian Staskawicz, University of California, Berkeley, California
Jack Ward Thomas, University of Montana, Missoula, Montana
James Tumlinson, United States Department of Agriculture, Agricultural Research Service, Gainesville, Florida
B.L. Turner, Clark University, Worcester, Massachusetts

Staff

Charlotte Kirk Baer, Director
Stephanie Padgham, Senior Project Assistant

BOARD ON ENVIRONMENTAL STUDIES AND TOXICOLOGY

Gordon Orians *(Chair)*, University of Washington, Seattle, Washington
John Doull *(Vice Chair)*, University of Kansas Medical Center, Kansas City, Missouri
David Allen, University of Texas, Austin, Texas
Thomas Burke, Johns Hopkins University, Baltimore, Maryland
Judith C. Chow, Desert Research Institute, Reno, Nevada
Christopher B. Field, Carnegie Institute of Washington, Stanford, California
William H. Glaze, University of North Carolina, Chapel Hill, North Carolina
Sherri W. Goodman, Center for Naval Analyses, Alexandria, Virginia
Daniel S. Greenbaum, Health Effects Institute, Cambridge, Massachusetts
Rogene Henderson, Lovelace Respiratory Research Institute, Albuquerque, New Mexico
Carol Henry, American Chemistry Council, Arlington, Virginia
Robert Huggett, Michigan State University, East Lansing, Michigan
Barry L. Johnson, Emory University, Atlanta, Georgia
James H. Johnson, Howard University, Washington, D.C.
James A. MacMahon, Utah State University, Logan, Utah
Patrick V. O'Brien, Chevron Research and Technology, Richmond, California
Dorothy E. Patton, International Life Sciences Institute, Washington, D.C.
Ann Powers, Pace University School of Law, White Plains, New York
Louise M. Ryan, Harvard University, Boston, Massachusetts
Jonathan M. Samet, Johns Hopkins University, Baltimore, Maryland
Kirk Smith, University of California, Berkeley, California
Lisa Speer, Natural Resources Defense Council, New York, New York
G. David Tilman, University of Minnesota, St. Paul, Minnesota
Chris G. Whipple, Environ Incorporated, Emeryville, California
Lauren A. Zeise, California Environmental Protection Agency, Oakland, California

Staff

James J. Reisa, Director
Ray Wassel, Program Director
Mimi Anderson, Senior Project Assistant

Acknowledgments

This report represents the integrated efforts of many individuals. The committee thanks all those who shared their insights and knowledge to bring the document to fruition. We also thank all those who provided information at our public meetings and who participated in our public sessions.

During the course of its deliberations, the committee sought assistance from several people who gave generously of their time to provide advice and information that were considered in its deliberations. Special thanks are due the following:

JOHN ALBERTSON, Duke University, Durham, North Carolina
C. RICHARD AMERMAN, United States Department of Agriculture, Beltsville, Maryland
BOB BOTTCHER (Deceased), North Carolina State University, Raleigh, North Carolina
GARTH BOYD, Murphy-Brown LLC, Warsaw, North Carolina
LEONARD BULL, Animal and Poultry Waste Center, Raleigh, North Carolina
TOM CHRISTENSEN, United States Department of Agriculture, Beltsville, Maryland
JOHN D. CRENSHAW, Eastern Research Group, Research Triangle Park, North Carolina
TONY DELANY, National Center for Atmospheric Research, Boulder, Colorado
RALPH ERNST, University of California, Davis, California
MICHAEL FITZGIBBON, California Environmental Protection Agency, Sacramento, California

Eric Gonder, Goldsboro Milling Company, Goldsboro, North Carolina
Alex Guenther, National Center for Atmospheric Research, Boulder, Colorado
Ellen Hankes, Environmental Management Solutions, LLC, Des Moines, Iowa
Lowry Harper, United States Department of Agriculture, Watkinsville, Georgia
Bruce Harris, United States Environmental Protection Agency, Research Triangle Park, North Carolina
Tom Horst, National Center for Atmospheric Research, Boulder, Colorado
Donald Johnson, Colorado State University, Fort Collins, Colorado
Renee Johnson, United States Environmental Protection Agency, Washington, D.C.
Ray Knighton, United States Department of Agriculture, Beltsville, Maryland
Gary Margheim, United States Department of Agriculture, Washington, D.C.
John H. Martin, Jr., Hall Associates, Dover, Delaware
F. Robert McGregor, Water and Waste Engineering, Incorporated, Denver, Colorado
Deanne Meyer, University of California, Davis, California
Bob Moser, ConAgra Beef, Kersey, Colorado
Daniel Murphy, National Oceanic and Atmospheric Administration, Boulder, Colorado
Brent Newell, California Rural Legal Assistance Foundation, Sacramento, California
Roy Oommen, Eastern Research Group, Research Triangle Park, North Carolina
Joseph Rudek, Environmental Defense, Raleigh, North Carolina
Gary Saunders, North Carolina Department of Environment and Natural Resources, Raleigh, North Carolina
Susan Schiffman, Duke University, Durham, North Carolina
Sally Shaver, United States Environmental Protection Agency, Research Triangle Park, North Carolina
Mark Sobsey, University of North Carolina, Chapel Hill, North Carolina
John Sweeten, Texas A&M University, Amarillo, Texas
David Townsend, Premium Standard Farms Research and Development, Kansas City, Missouri
Randy Waite, United States Environmental Protection Agency, Research Triangle Park, North Carolina
John T. Walker, United States Environmental Protection Agency, Research Triangle Park, North Carolina

The committee is grateful to members of the National Research Council staff who worked diligently to maintain progress and quality in its work.

The report has been reviewed in draft form by individuals chosen for their diverse perspectives and technical expertise, in accordance with procedures approved by the National Research Council's Report Review Committee. The purpose of this independent review is to provide candid and critical comments that will assist the institution in making its published report as sound as possible and to ensure that the report meets institutional standards for objectivity, evidence, and responsiveness to the study charge. The review comments and draft manuscript remain confidential to protect the integrity of the deliberative process. We wish to thank the following individuals for their review of this report:

David T. Allen, The University of Texas, Austin, Texas
William Battye, EC/R Incorporated, Chapel Hill, North Carolina
Van C. Bowersox, Illinois State Water Survey, Champaign, Illinois
Ellis B. Cowling, North Carolina State University, Raleigh, North Carolina
Danny G. Fox, Cornell University, Ithaca, New York
Rogene Henderson, National Environmental Respiratory Center, Albuquerque, New Mexico
Kristen A. Johnson, Washington State University, Pullman, Washington
Deanne Meyer, University of California, Davis, California
George Mount, Washington State University, Pullman, Washington
Roger A. Pielke, Colorado State University, Fort Collins, Colorado
Wendy J. Powers, Iowa State University, Ames, Iowa
Joseph Rudek, Environmental Defense, Raleigh, North Carolina
Margot Rudstrom, University of Minnesota, Morris, Minnesota

Although the reviewers listed above have provided many constructive comments and suggestions, they were not asked to endorse the conclusions or recommendations, nor did they see the final draft of the report before its release. The review of this report was overseen by Bob Frosch, Harvard University, Cambridge, Massachusetts, and Albert Heber, Purdue University, West Lafayette, Indiana. Appointed by the National Research Council, they were responsible for making certain that an independent examination of this report was carried out in accordance with institutional procedures and that all review comments were carefully considered. Responsibility for the final content of this report rests entirely with the authoring committee and the institution.

Preface

The increasing concentration of food production—meat, eggs, milk—from animals in very large feeding operations has focused public attention on associated environmental issues. These include the effects of air emissions, especially those that come from the large quantities of manure produced by the animals. While concern has mounted, research to provide the basic information needed for effective regulation and management of these emissions has languished.

This report, prepared by a committee appointed by the National Research Council, proposes two major ways to improve information and the nation's ability to deal with the effects of these emissions. One is to change the way in which the rates and fate of air emissions are estimated and tracked. The proposal would replace the current "emission factor" approach with a "process-based modeling" approach. This can, if pursued vigorously, enhance both regulation and management of air emissions in the next two to five years.

The other proposal is for a research program that views air emissions as one part of the overall system of producing food from animal feeding operations with the goal of eliminating the release of unwanted emissions into the environment. This "systems-based" proposal, if also pursued vigorously, would lead to fundamentally changed practices at animal feeding operations. The net result would be continued food production with greatly reduced adverse environmental effects.

The 16-person committee that produced this report and an earlier interim report worked hard and well. The time allowed for producing the two reports was short, but committee members found time in their schedules to address what each sees as an important issue that needs attention. The project staff at the Board on Agriculture and Natural Resources, Jamie Jonker, study director, and Tanja Pilzak, research assistant, and the Board on Environmental Studies and Toxicology, Chad Tolman, program officer, deserve special thanks for their long hours of very effective work. An informal editorial subcommittee that handled reviewer comments and provided enormous help throughout also deserves special thanks. The members were Chair Perry Hagenstein, Vice Chair Bob Flocchini, Jim Galloway, Rick Kohn, and, for the interim report, Wayne Robarge.

Perry Hagenstein, *Chair*
Robert Flocchini, *Vice Chair*
Committee on Air Emissions from
Animal Feeding Operations

Contents

Tables, Figures, and Boxes

TABLES

FIGURES

BOXES

Executive Summary

Public concerns about the environmental effects and, to a lesser extent, the possible health effects of air emissions from animal feeding operations (AFOs, see Appendix B) have grown with the increasing size and geographic concentration of these operations. This intensification has been driven by the economics of domestic and export markets for meat, poultry, milk, and eggs. Public concerns have also grown as the population, both exurbanites and expanding urban centers, have moved into what had been largely rural farming areas. Objectionable odors from AFOs are a significant concern not only to the new residents in these areas, but also to many long-time residents.

Prompted by legislation, especially the Clean Air Act (CAA), as well as by public concerns, the U.S. Environmental Protection Agency (EPA) has been considering what information is needed to define and support feasible regulation of air emissions from AFOs. At the same time, the U.S. Department of Agriculture (USDA) has been using its authority to aid farmers in mitigating the effects of air emissions with modified agricultural practices. Acting jointly, these two agencies asked the Board on Agriculture and Natural Resources (BANR) to evaluate the scientific information needed to address these issues. A 16-person ad hoc committee was appointed, the Committee on Air Emissions from Animal Feeding Operations, which has been guided by a Statement of Task that was agreed upon by the National Academies and the sponsoring agencies (Appendix A).

The Statement of Task directed the committee to

- review and evaluate the scientific basis for estimating the emissions to the atmosphere of various specified substances from confined livestock and poultry operations;

- review the characteristics of the agricultural animal industries, methods for measuring and estimating air emissions, and potential best management practices for mitigating emissions;
- evaluate confined animal feeding production systems in terms of biologic systems; and
- identify critical short- and long-term research needs and recommend methodologic and modeling approaches for estimating and measuring air emissions and potential mitigation technologies.

Making scientifically credible estimates of air emissions from AFOs is complicated by various factors that affect the amounts and dispersion of emissions in the atmosphere. Such factors include the kinds and numbers of animals involved, their diets and housing, the management of their manure (feces and urine, which may also include litter or bedding materials), topography, climatic and weather conditions, and actions taken to mitigate the emissions and their effects. Estimates of emissions generated for one set of conditions or for one type of AFO may not translate readily to others.

Accurate estimation of air emissions from AFOs is needed to gauge their possible adverse impacts and the subsequent implementation of control measures. For example, increasing pressure is being placed on EPA to address these emissions through the Clean Air Act and other federal laws and regulations. EPA is under court order to establish new water quality rules for AFOs by December 2002. The need to understand the relationship between actions to mitigate the effects of manure management on water quality and its related effects on air quality prompted EPA to ask for an interim report several months in advance of this final report. The committee's findings in the interim report (Box ES-1) are encompassed and extended by the findings and recommendations in this report.

The contents, including the findings and recommendations, of this report represent the consensus views of the committee and have been formally reviewed in accordance with National Research Council procedures. In addressing its Statement of Task, the committee has come to consensus on 13 major findings, each accompanied by one or more related recommendations. The basis of these findings is discussed more extensively in the body of the report.

FINDINGS AND RECOMMENDATIONS

Animal Units

EPA defines animal units differently than USDA. An EPA animal unit is equal to 1.0 slaughter and feeder cattle, 0.7 mature dairy cows, 2.5 pigs weighing more than 55 pounds, 10 sheep or lambs, and 0.5 horses. USDA defines animal unit as 454 kg (1000 pounds) of animal live weight regardless of species. A consistent basis for defining animal unit will decrease confusion that may exist be-

BOX ES-1 Findings from the Interim Report (NRC, 2002a)

Finding 1. Proposed EPA regulations aimed at improving water quality may affect rates and distribution of air emissions from animal feeding operations.

Finding 2. In order to understand health and environmental impacts on a variety of spatial scales, estimates of air emissions from AFOs at the individual farm level, and their dependence on management practices, are needed to characterize annual emission inventories for some pollutants and transient downwind spatial distributions and concentrations for others.

Finding 3. Direct measurements of air emissions at all AFOs are not feasible. Nevertheless, measurements on a statistically representative subset of AFOs are needed and will require additional resources to conduct.

Finding 4. Characterizing feeding operations in terms of their components (e.g., model farms) may be a plausible approach for developing estimates of air emissions from individual farms or regions as long as the components or factors chosen to characterize the feeding operation are appropriate. The method may not be useful for estimating acute health effects, which normally depend on human exposure to some concentration of toxic or infectious substance for short periods of time.

Finding 5. Reasonably accurate estimates of air emissions from AFOs at the individual farm level require defined relationships between air emissions and various factors. Depending on the character of the AFOs in question, these factors may include animal types, nutrient inputs, manure handling practices, output of animal products, management of feeding operations, confinement conditions, physical characteristics of the site, and climate and weather conditions.

Finding 6. The model farm construct as described by EPA (2001a) cannot be supported because of weaknesses in the data needed to implement it.

Finding 7. The model farm construct used by EPA (2001a) cannot be supported for estimating either the annual amounts or the temporal distributions of air emissions on an individual farm, subregional, or regional basis because the way in which it characterizes feeding operations is inadequate.

Finding 8. A process-based model farm approach that incorporates "mass balance" constraints for some of the emitted substances of concern, in conjunction with estimated emission factors for other substances, may be a useful alternative to the model farm construct defined by EPA (2001a).

SOURCE: NRC (2002a).

cause of the differing definitions. The process-based model described in this report is better suited for using a continuous variable (e.g., 500-kg live weight) than a discrete variable (e.g., 1 dairy cow).

FINDING 1. Much confusion exists about the use of the term "animal unit" because EPA and USDA define animal unit differently.

RECOMMENDATION: Both EPA and USDA should agree to define animal unit in terms of animal live weight rather than an arbitrary definition of animal unit.

Spatial Distribution of Effects

The various substances that together make up the total air emissions from animal feeding operations differ in quantity, the potential severity of their effects, and the spatial distribution of these effects. Ammonia, whose environmental impacts are reasonably well understood, has relevant impacts that have to be addressed at regional, national, and global scales. On the other hand, odor, whose composition is not well known in scientific terms and whose impacts on the public are difficult to judge, is important mainly at a very local level.

Table ES-1, which supports and elaborates Finding 2 below, represents the reasoned judgment of the committee on the relative importance of each substance at the relevant spatial scales strictly for emissions from AFOs. For example, volatile organic compounds (VOCs) play an important role in tropospheric ozone formation, yet such emissions from AFOs are likely to be insignificant compared to other sources in most areas.

FINDING 2. Air emissions from animal feeding operations are of varying concern at different spatial scales, as shown in Table ES-1.

RECOMMENDATION: These differing effects, concentrations, and spatial distributions lead to a logical plan of action for establishing research priorities to provide detailed scientific information on the contributions of AFO emissions to potential effects and the subsequent implementation of control measures. USDA and EPA should first focus their efforts on the measurement and control of those emissions of major concern.

Measurement Protocols and Control Technologies

Achieving the overall goal of decreasing the adverse impacts of air emissions from AFOs will require attention to the differences in the character of the various emissions (e.g., their persistence in the atmosphere), in the way they are dis-

TABLE ES-1 Committee's Scientific Evaluation of the Potential Importance[a] of AFO Emissions at Different Spatial Scales

Emissions	Global, National, and Regional	Local—Property Line or Nearest Dwelling	Primary Effects of Concern
NH_3	Major[a]	Minor	Atmospheric deposition, haze
N_2O	Significant	Insignificant	Global climate change
NO_x	Significant	Minor	Haze, atmospheric deposition, smog
CH_4	Significant	Insignificant	Global climate change
VOCs[b]	Insignificant	Minor	Quality of human life
H_2S	Insignificant	Significant	Quality of human life
PM10[c]	Insignificant	Significant	Haze
PM2.5[c]	Insignificant	Significant	Health, haze
Odor	Insignificant	Major	Quality of human life

[a]Relative importance of emissions from AFOs at spatial scales based on committees' informed judgment on known or potential impacts from AFOs. Rank order from high to low importance is major, significant, minor, and insignificant. While AFOs may not play an important role for some of these, emissions from other sources alone or in aggregate may have different rankings. For example VOCs and NO_x play important roles in the formation of tropospheric ozone; however, the role of AFOs is likely to be insignificant compared to other sources.

[b]Volatile organic compounds.

[c]Particulate matter. PM10 and PM2.5 include particles with aerodynamic equivalent diameters up to 10 and 2.5 μm, respectively.

persed, in their environmental effects, and in the effectiveness of various control and management strategies. As noted above, it will also require attention to priorities based on the geographic scale at which impacts are of greatest concern. The local scale is considered the AFO boundary or nearest occupied dwelling. The regional scale may be as small as a single topographic land feature (e.g., a stream valley) or as large as a multistate airshed.

FINDING 3. Measurement protocols, control strategies, and management techniques must be emission and scale specific.

RECOMMENDATIONS:

- **For air emissions important on a global or national scale (i.e., ammonia and the greenhouse gases methane [CH_4] and nitrous oxide [N_2O]), the aim is to control emissions per unit of production (kilograms of food produced) rather than emissions per farm. Where the environmental and health benefits outweigh the costs of mitigation it is important to decrease aggregate emissions. In some geographic regions,**

aggregate emission goals may limit the number of animals produced in those regions.

- **For air emissions important on a local scale (hydrogen sulfide [H_2S], particulate matter [PM], and odor), the aim is to control ambient concentrations at the farm boundary and/or nearest occupied dwelling. Standards applicable to the farm boundary and/or nearest occupied dwelling must be developed.**
- **Monitoring should be conducted to measure concentrations of air pollutants of possible health concern at times when they are likely to be highest and in places where the densities of animals and humans, and typical meteorological conditions, are likely to result in the highest degree of human exposure.**

Current Best Management Plans

As noted in the committee's interim report, available estimates of emission factors, rates, and concentrations are sufficiently uncertain that they provide a poor basis for regulating or managing air emissions from AFOs. Nevertheless, some best management practices to mitigate the adverse effects of air emissions appear at face value to warrant their use, even as new information on mitigation and best management practices is being developed. Although the committee favors a strong focus on research to develop needed new information, the use of clearly effective measures should be encouraged while new information is being developed.

FINDING 4. There is a general paucity of credible scientific information on the effects of mitigation technologies on concentrations, rates, and fates of air emissions from AFOs. However, the implementation of technically and economically feasible management practices (e.g., manure incorporation into soil) designed to decrease emissions should not be delayed.

RECOMMENDATION: Best management practices (BMPs) aimed at mitigating AFO air emissions should continue to be improved and applied as new information is developed on the character, amount, and dispersion of these air emissions, and on their health and environmental effects. A systems analysis should include impacts of a BMP on other parts of the entire system.

Odors

Odors associated with AFO emissions are often regulated in response to nuisance complaints rather than demonstrated health effects. The measurement of odor concentrations downwind from AFOs is based on olfactometers that relate odor strength to a standard (usually *n*-butanol) or uses the judgment of panels of experts trained to distinguish odor strengths. While standardized terminology and measures have been developed in Europe, a similar effort has not yet occurred in the scientific community in the United States.

Odors continue to be a problem with AFOs at the local level. Continuing research into the constituents of odor with a goal of providing a basis for scientific agreement for standards is needed.

FINDING 5. Standardized methodologies for odor measurement have not been adopted in the United States.

RECOMMENDATIONS:

- **Standardized methodology should be developed in the United States for objective measurement techniques of odors to correspond to subjective human response.**
- **A standardized unit of measurement of odor concentration should be adopted in the United States.**

Dispersion Modeling

FINDING 6. The complexities of various kinds of air emissions and the temporal and spatial scales of their distribution make direct measurement at the individual farm level impractical other than in a research setting. Research into the application of advanced three-dimensional modeling techniques accounting for transport over complex terrain under thermodynamically stable and unstable planetary boundary layer (PBL) conditions offers good possibilities for improving emissions estimates from AFOs.

RECOMMENDATION: EPA should develop and carry out one or more intensive field campaigns to evaluate the extent to which ambient atmospheric concentrations of the various species of interest are consistent with estimated emissions and to understand how transport and chemical dynamics shape the local and regional distribution of these species.

Measurement Protocols

Accurate measurement of air emissions is dependent on the availability and use of protocols that are both technically sound and practical for use in the field, as well as the laboratory. Such measurement protocols are available for measuring nitrous oxide and nitric oxide (NO). Improved measurement protocols are needed for other substances. Particulate matter, odor, and volatile organic compounds are important emissions at the local level, but pose some special problems because their constituents and emission rates vary widely among AFOs and their locations.

FINDING 7. Scientifically sound and practical protocols for measuring air concentrations, emission rates, and fates are needed for the various elements (nitrogen, carbon, sulfur), compounds (e.g., ammonia [NH_3], CH_4, H_2S), and particulate matter.

RECOMMENDATIONS:

- **Reliable and accurate calibration standards should be developed, particularly for ammonia.**
- **Standardized sampling and compositional analysis techniques should be provided for PM, odor, and their individual components.**
- **The accuracy and precision of analytical techniques for ammonia and odor should be determined, including intercomparisons on controlled (i.e., synthetic) and ambient air.**

Emission Factors

The "emission factor" approach for estimating air emissions is based on measuring emissions from a set of defined AFOs to obtain an "average" emission per unit (e.g., per animal unit or per unit of production). These emission factors can then be used to estimate emissions for other AFOs by multiplying the emission factor by the number of observed units to which the average applies. As noted in the committee's interim report, the existing emission factors for AFOs are generally inadequate because of the limited number of measurements on which they are based, as well as the limited generality of the models for which the emission factors have been developed (see Appendix L). Improving existing emission factors to the point where they could provide scientifically credible estimates of either emission rates or concentrations would require major efforts in getting sufficient observations to characterize the variability among and within AFOs.

The committee (in Finding 9 and Chapter 5) suggests that an alternative approach for estimating emissions, a process-based modeling approach, can provide more useful estimates for most of the air emission substances of concern. Particulate matter is the main exception and may require additional efforts to improve emission factors. Allocation of overall resources for improving and evaluating emission estimates should focus on the committee's recommended process-based modeling approach for all emissions mentioned, except for particulate matter.

FINDING 8. Estimating air emissions from AFOs by multiplying the number of animal units by existing emission factors is not appropriate for most substances.

RECOMMENDATION: The science for estimating air emissions from individual AFOs should be strengthened to provide a broadly recognized and acceptable basis for regulations and management programs aimed at mitigating the effects of air emissions.

Process-Based Model

To counter the tendency to consider only on-farm inputs and outputs from AFOs, to ensure more accurate accounting of the flows of chemicals and other air emission substances from the operation, and to provide a "mass balance" control for the total flow of inputs to and outputs from the operation, the committee recommends a "process-based modeling" approach for estimating air emissions.

The process-based modeling approach can be used to estimate the flows of elements (nitrogen, sulfur, carbon) and of compounds containing these elements. The committee believes, with some reservations, that this approach might be used for estimating odor emissions. The only substance of direct concern to the committee for which this approach may not be well suited is particulate matter.

This approach involves the specification of mathematical models that describe the movement of various substances of interest at each major stage of the process of producing livestock products: movement into the next stage, movement in various forms to the environment, and ultimately movement into products used by humans. Mass balance constraints serve as a check on the whole system to ensure that estimates of movements of substances out of the system do not exceed the amounts available within the system.

FINDING 9. Use of process-based modeling will help provide scientifically sound estimates of air emissions from AFOs for use in regulatory and management programs.

RECOMMENDATIONS:

- **EPA and USDA should use process-based mathematical models with mass balance constraints for nitrogen-containing compounds, methane, and hydrogen sulfide to identify, estimate, and guide management changes that decrease emissions for regulatory and management programs.**
- **EPA and USDA should investigate the potential use of a process-based model to estimate mass emissions of odorous compounds and potential management strategies to decrease their impacts.**

- **EPA and USDA should commit resources and adapt current or adopt new programs to fill identified gaps in research to improve mathematical process-based models to increase the accuracy and simplicity of measuring and predicting emissions from AFOs (see short-term and long-term research recommendations).**

Systems Analysis

The emission factor approaches in current use focus on the "on-farm" inputs to and outputs from an AFO. This ignores the potential environmental effects associated with "off-farm" production of feed and other materials used in an AFO. Since some of the feed for typical AFOs is imported from other farms, and a portion of the manure is often exported from the AFO for use on other farms (some regional and species differences exist), restricting consideration of inputs and outputs to a single AFO may not completely represent the full environmental effects of the operation. A "systems approach" that considers both the on-farm and the off-farm inputs and outputs would provide a more accurate description of overall impacts.

FINDING 10. A systems approach, which integrates animal and crop production systems both on and off (imported feeds and exported manure) the AFO, is necessary to evaluate air emissions from the total animal production system.

RECOMMENDATION: Regulatory and management programs to decrease air emissions should be integrated with other environmental (e.g., water quality) and economic considerations to optimize public benefits.

Nitrogen Emissions

Because of its potential environmental impacts at regional, national, and global scales, instituting control strategies for nitrogen emissions should be assigned high priority. Sufficient information is currently available to do this at all geographic scales.

FINDING 11. Nitrogen emissions from AFOs and total animal production systems are substantial and can be quantified and documented on an annual basis. Measurements and estimates of individual nitrogen species components (i.e., NH_3, molecular nitrogen [N_2], N_2O, and NO) should be made in the context of total nitrogen losses.

RECOMMENDATION: Control strategies aimed at decreasing emissions of reactive nitrogen compounds (Nr) from total animal production

systems should be designed and implemented now. These strategies can include both performance standards based on individual farm calculations of nitrogen balance and technology standards to decrease total system emissions of reactive nitrogen compounds by quantifiable amounts.

Research

The two major federal agencies with regulatory or management responsibilities relative to air emissions from AFOs are EPA and USDA. Each of these agencies also has research responsibilities in support of its action programs—responsibilities that are typically serviced through "in-house" research staffs. Close cooperation is needed between the two agencies in setting and supporting research priorities relative to air emissions. Inputs and participation from the full range of state, private, and research institutions with relevant interests are needed to ensure that concerns about air emissions are addressed with the full complement of needed expertise.

The importance of food production from AFOs, coupled with the potential environmental effects from air emissions, demands substantial research efforts in both the short and the long term. These issues will not be resolved without addressing the appropriate funding of these efforts. Current allocations of funding aimed at AFO air emissions are not adequate or appropriate in view of the amount of concern about these emissions and the recent growth in AFO livestock production.

Research in the short term (four to five years) can significantly improve the capability of the process-based modeling approach for estimating air emissions. A long-term (20-30 years) research program that encompasses overall impacts of animal production on the environment, as proposed here, can have even more substantial results in decreasing overall impacts on the environment, while sustaining production at a high level.

FINDING 12. USDA and EPA have not devoted the necessary financial or technical resources to estimate air emissions from AFOs and develop mitigation technologies. The scientific knowledge needed to guide regulatory and management actions requires close cooperation between the major federal agencies (EPA and USDA), the states, industry and environmental interests, and the research community, including universities.

RECOMMENDATIONS:

- **EPA and USDA should cooperate in forming a continuing research coordinating council (1) to develop a national research agenda on issues related to air emissions from AFOs in the context of animal production systems and (2) to provide continuing oversight on the imple-**

mentation of this agenda. This council should include representatives of EPA and USDA, the research community, and other relevant interests. It should have authority to advise on research priorities and funding.

- **Exchanges of personnel among the relevant agencies should be promoted to encourage efficient use of personnel, broadened understanding of the issues, and enhanced cooperation among the agencies.**
- **For the short term, USDA and EPA should initiate and conduct a coordinated research program designed to produce a scientifically sound basis for measuring and estimating air emissions from AFOs on local, regional, and national scales.**
- **For the long term, USDA, EPA, and other relevant organizations should conduct coordinated research to determine which emissions (to water and air) from animal production systems are most harmful to the environment and human health, and to develop technologies that decrease their releases into the environment. The overall research program should include research to optimize inputs to AFOs, optimize recycling of materials, and significantly decrease releases to the environment.**

The reality of budget constraints in allocating research funds to address problems of air emissions requires a careful weighing of several factors, including those that affect both the implementation costs and the societal benefits. Finding 2 proposes a way of ranking both action and research opportunities among the emission substances based on amounts of concern or impacts and geographic scale of impacts. A more complete listing of factors is needed for setting both short- and long-term research priorities and for allocating research funds.

FINDING 13. Setting priorities for both short- and long-term research on estimating air emission rates, concentrations, and dispersion requires weighing the potential severity of adverse impacts, the extent of current scientific knowledge about them, the potential for advancing scientific knowledge, and the potential for developing successful mitigation and control strategies.

RECOMMENDATIONS:

- **Short-term research priorities should improve estimates of emissions from individual AFOs including the effects of different control technologies:**
- **Priority research for emissions important on a local scale should be conducted on odor, PM, and H_2S (also see Finding 2).**
- **Priority research for emissions important on regional, national, and**

global scales should be conducted on ammonia, N_2O, and methane (also see Finding 2).

- **Long-term research priorities should improve understanding of animal production systems and lead to development of new control technologies.**

SUMMARY

These findings and recommendations, taken together, point to two major changes in direction for improving the basic information needed for dealing with the adverse effects of air emissions from AFOs. One is to replace the current emission factor approach for estimating and tracking the rates and fates of air emissions using a process-based modeling approach with mass balance constraints. The second is to initiate a substantial long-term research program on the overall system of producing food from animal feeding operations with the goal of eliminating the release of undesirable air and other emissions into the environment.

Facing the need for defensible information on air emissions from AFOs, in a timely manner, is a major challenge for EPA and USDA. Neither has yet addressed the need for this information in defining high-priority research programs. Each has pursued its regulatory and farm management programs under the assumption that the best currently available information can be used to implement its program goals.

The scope and complexity of the information needed by these agencies, as well as the potential environmental impacts of air emissions from AFOs, require a concentrated, focused, and well-funded research effort. Such an effort is described in this report.

1

Introduction

BASIS FOR THIS REPORT

Increases in the size and geographical concentration (see Appendix K) of animal feeding operations (AFOs; see Appendix B) and growing concerns with emissions from them appear to be leading toward regulation or other means to mitigate their air emissions. Recognizing the need for solid scientific information on which to base regulatory or other program decisions, the U.S. Environmental Protection Agency (EPA) and the U.S. Department of Agriculture (USDA) asked the Board on Agriculture and Natural Resources to evaluate the scientific basis for estimating various kinds of air emissions from AFOs. The specific requirements to guide this study were dictated by the committee's Statement of Task (Appendix A). A 16-person committee was appointed with expertise in various relevant disciplines (see "About the Authors" section) to conduct the study.

The policy and program issues connected to air emissions from AFOs are multifaceted. The agencies that sponsored and funded this study (EPA and USDA) have direct and indirect program interests. Under the Clean Air Act (CAA) and the Clean Water Act (CWA), EPA is responsible for defining regulatory programs through the states to improve and maintain the nation's air and water quality. USDA sponsors programs to provide farmers and other landowners with technical and financial assistance to adopt practices that will improve the environmental quality of land and related air and water resources. Both agencies have research programs aimed at providing scientific information necessary for pursuing their program goals. In addition, lawmakers and those making policy decisions at all levels of government require solid scientific information to carry out their tasks. This report evaluates the availability of this information and proposes ways to acquire it.

The research programs supported by EPA and USDA are obvious candidates for providing much of the needed information. USDA has by far the largest overall research program of the two agencies. It has in-house capabilities (Agricultural Research Service, Economic Research Service, and National Agricultural Statistics Service) and helps support an extensive extramural research program conducted through state universities and a system of agricultural research stations. Total funding for these programs is about five times that of EPA's program, which is conducted largely through a system of EPA research labs. Only small parts of these programs, however, are devoted to research related to air emissions.

The findings and recommendations in this report are aimed in large part at the leaders and scientists of the EPA and USDA research programs, but they are also aimed at the entire community interested in addressing the issues posed by the adverse effects of air emissions from animal feeding operations. This includes leaders in the scientific research community, agriculture in general, environmental interests, people affected by the emissions, and the farmers who ultimately have to deal directly with their causes.

CONCERNS WITH AIR EMISSIONS

The EPA and USDA have asked the committee to address the issues relating to the substances shown in Table 1-1. The committee added nitric oxide (NO) because it is produced by AFOs and their associated grain production and manure disposal, and because it can have significant environmental effects. As Table 1-1 indicates, the substances of concern vary in their classifications as air pollutants. They also vary in the severity and scale of their effects.

The issue that most often brings air emissions to the attention of public officials is the frequency of complaints about strong and objectionable odors voiced by neighbors of large feeding operations. Additionally, particulate matter may blow from farms to nearby residences and trouble residents because of actual or perceived health effects. Equally important are the various substances in air emissions that contribute to environmental degradation, such as eutrophication of water bodies (caused by reactive nitrogen compounds) or climate change (induced by the greenhouse gases methane and nitrous oxide). The committee believes that these concerns warrant serious attention to determine the effects of AFOs and to mitigate their detrimental air emissions.

THE INTERIM REPORT

As part of its charge, EPA asked that the committee provide an interim report in the spring of 2002 to give it an early indication of findings that would help in planning regulations to decrease impacts of AFOs on water quality. In particular, EPA was concerned that possible actions to improve water quality might have an

TABLE 1-1 Substances in AFO Emissions That the Committee Was Tasked to Address and Their Respective Classifications[a]

Species	Criteria Pollutant	Hazardous Air Pollutant (HAP)	Greenhouse Gas	Regulated Air Pollutant
NH_3[b]	—	—	—	X
N_2O[c]	—	—	X	—
NO_x	X	—	—	X
CH_4	—	—	X	—
VOCs[d]	—	X[e]	—	X[f]
H_2S[g]	—	—	—	X
PM (TSP)[h]	—	—	—	X
PM10	X	—	—	X
PM2.5	X	—	—	X
Odor[i]	—	—	—	X

[a]See Appendix B for definitions and Chapter 6 for regulations.
[b]Ammonia is not a criteria pollutant but is a precursor for secondary PM2.5, which is a criteria pollutant.
[c]Nitrous oxide is not a precursor for the formation of tropospheric ozone, but is a greenhouse gas. It is not considered to be part of NO_x (the sum of NO and NO_2), which contributes to formation of ozone, a criteria pollutant.
[d]Volatile organic compounds (VOCs), sometimes referred to as reactive organic gases (ROGs), contribute to the formation of ozone, a criteria pollutant.
[e]Some, but not all, VOCs are listed as Hazardous Air Pollutants (HAPs).
[f]Some VOCs are regulated as HAPs, and some are regulated as ozone precursors.
[g]Hydrogen sulfide is not listed as a criteria pollutant or a hazardous air pollutant (HAP). However, it is a regulated pollutant because it is listed as having a New Source Performance Standard (NSPS). It may be added to the HAPs list in the near future.
[h]Particulate matter. Prior to 1987, PM was a criteria pollutant and regulated as total suspended particulate (TSP). Currently, the PM fractions listed as criteria pollutants are PM10 and PM2.5. However, TSP emissions are regulated in some states.
[i]Odor is a regulated pollutant in some states. State air pollution regulatory agencies regulate it based on a nuisance standard.

adverse impact on air emissions. The committee was also asked to assess the approach for estimating air emissions from AFOs presented in a draft contract report to EPA *Emissions from Animal Feeding Operations* (EPA, 2001a). The committee's interim report provided EPA with findings on the following: identification of the scientific criteria needed to ensure that air emission rates are reasonable, the basis for these criteria as documented in the scientific literature, and the uncertainties associated with them. The interim report was reviewed in accordance with National Research Council procedures. It responded directly to a series of questions posed by EPA:

- What are the scientific criteria needed to ensure that reasonably appropriate estimates of emissions are obtained? What are the strengths, weaknesses, and gaps of published methods to measure specific emissions and develop emission factors that are published in the scientific literature? How should the variability due to regional differences, daily and seasonal changes, animal life stage, and different management approaches be characterized? How should the statistical uncertainty in emissions measurements and emissions factors be characterized in the scientific literature?

- Are the emission estimation approaches described in the EPA report *Emissions from Animal Feeding Operations* (EPA, 2001a) appropriate? If not, how should industry characteristics and emission mitigation techniques be characterized? Should model farms be used to represent the industry? If so, how? What substances should be characterized and how can inherent fluctuations be accounted for? What components of manure should be included in the estimation approaches (e.g., nitrogen, sulfur, volatile solids)? What additional emission mitigation technologies and management practices should be considered?
- What criteria, including capital costs, operating costs, and technical feasibility, are needed to develop and assess the effectiveness of emission mitigation techniques and best management practices?

Responses to those questions are summarized in eight findings along with a brief discussion of each finding in the committee's interim report (NRC, 2002a; Box 1-1). These interim report findings provide a foundation for the findings and recommendations in this report, which points to the limitations of currently available information on air emissions from AFOs. It also points to the need for new approaches to make such estimates, and describes further research needed to support regulatory and management programs aimed at decreasing air emissions.

SCALE AND IMPACT OF EMISSIONS FROM ANIMAL FEEDING OPERATIONS

The scope of the issues arising from AFO air emissions is large. A large fraction of the crops grown in the United States are fed to domesticated animals to produce meat, milk, and eggs for human consumption. As animal populations have grown in some locations and become more concentrated on larger farms, and as humans leave urban areas, concern has increased because of possible adverse effects on human health and the environment.

Between 1982 and 1997, the number of animal feeding operations in the United States decreased by 51 percent, while livestock production increased 10 percent (Gollehon et al., 2001). In some areas, even greater changes in concentration have occurred (G. Saunders, North Carolina Department of Environment and Natural Resources, personal communication, 2002). As AFOs have increased in size and geographical concentration (see Appendix K), the potential health and

BOX 1-1 Findings and Discussion from the Interim Report

Finding 1. Proposed EPA regulations aimed at improving water quality may affect rates and distributions of air emissions from animal feeding operations.

Discussion: Regulations aimed at protecting water quality would probably affect manure management at the farm level, especially since they might affect the use of lagoons and the application of manure on cropland or forests. For example, the proposed water regulations may mandate nitrogen (N) or phosphorus (P) based comprehensive nutrient management plans (CNMPs). AFOs could be limited in the amount of manure nitrogen and phosphorus that could be applied to cropland. If there is a low risk of phosphorus runoff as determined by a site analysis, farmers will be permitted to overapply phosphorus. However, they will still be prohibited from applying more nitrogen than recommended for crop production. Many AFOs (those currently without CNMPs) likely will have more manure than they can use on their own cropland, and manure export may be cost prohibitive. Thus, AFOs will have an incentive to use crops and management practices that employ applied nitrogen inefficiently (i.e., volatilize ammonia) to decrease the nitrogen remaining after storage or increase the nitrogen requirement for crop production. These practices may increase nitrogen volatilization to the air. The committee was not informed of specific regulatory actions being considered by EPA (beyond those addressed in the *Federal Register*) to meet its December 2002 deadline for proposing regulations under the Clean Water Act.

Finding 2. In order to understand health and environmental impacts on a variety of spatial scales, estimates of air emissions from AFOs at the individual farm level, and their dependence on management practices, are needed to characterize annual emission inventories for some pollutants and transient downwind spatial distributions and concentrations for others.

Discussion: Management practices (e.g., feeding, manure management, crop management) vary widely among individual farms. Estimates of emissions based on regional or other averages are unlikely to capture significant differences among farms that will be relevant for guiding emissions management practices aimed at decreasing their effects. Information on the spatial relationships among individual farms and the dispersion of air emissions from them is needed. Furthermore, developing methods to estimate emissions at the individual farm level was the stated objective of EPA's recent study (EPA, 2001a).

Finding 3. Direct measurements of air emissions at all AFOs are not feasible. Nevertheless, measurements on a statistically representative subset of AFOs are needed and will require additional resources to conduct.

Discussion: Although it is possible in a carefully designed research project to measure concentrations and airflows (e.g., building ventilation rates) to estimate air emissions and attribute them to individual AFOs, it is not practical to conduct such projects for more than a small fraction of AFOs. Direct measurements for sample farms will be needed in research programs designed to develop estimates of air emissions applicable to various situations.

Finding 4. Characterizing feeding operations in terms of their components (e.g., model farms) may be a plausible approach for developing estimates of air emissions from individual farms or regions as long as the components or factors chosen to characterize the feeding operation are appropriate. The method may not be useful for estimating acute health effects, which normally depend on human exposure to some concentration of toxic or infectious substance for short periods of time.
Discussion: The components or factors used to characterize feeding operations are chosen for their usefulness in explaining dependent variables, such as the mass of air emissions per unit of time. The emission factor method, which is based on the average amount of an emitted substance per unit of activity per year (e.g., metric tons of ammonia per thousand head of cattle per year), can be useful in estimating annual regional emissions inventories for some pollutants, provided that sufficient data of adequate quality are available for estimating the relationships.

Finding 5. Reasonably accurate estimates of air emissions from AFOs at the individual farm level require defined relationships between air emissions and various factors. Depending on the character of the AFOs in question, these factors may include animal types, nutrient inputs, manure handling practices, output of animal products, management of feeding operations, confinement conditions, physical characteristics of the site, and climate and weather conditions.
Discussion: The choice of independent variables used to make estimates of air emissions from AFOs will depend on the ability of the variables to account for variations in the estimates and on the degree of accuracy desired, based on valid measurements at the farm level. Past research indicates that some combination of the indicated variables is likely to be important for estimates of air emissions for the kinds of operations considered in this report. The specific choices will depend on the strength of the relationships for each kind of emission and each set of independent variables.

Finding 6. The model farm construct as described by EPA (2001a) cannot be supported because of weaknesses in the data needed to implement it.

continues

Discussion: Of the nearly 500 possible literature sources for estimating emissions factors identified for EPA (2001a), only 33 were found by the report's authors to be suitable for use in the model farm construct. The committee judged them to be insufficient for the intended use. The breadth in terms of kinds of animals, management practices, and geography in this model farm construct suggests that finding adequate information to define emission factors is unlikely to be fruitful at this time.

Finding 7. The model farm construct used by EPA (2001a) cannot be supported for estimating either the annual amounts or the temporal distributions of air emissions on an individual farm, subregional, or regional basis because the way in which it characterizes feeding operations is inadequate.
Discussion: Variations in many factors that could affect the annual amounts and temporal patterns of emissions from an individual AFO are not adequately considered by the EPA (2001a) model farm construct. The potential influences of geographic (e.g., topography and land use) and climatic differences, daily and seasonal weather cycles, animal life stages, management approaches (including manure management practices and feeding regimes), and differences in state regulations are not adequately considered. Furthermore, aggregating emissions from individual AFOs using the EPA (2001a; not a stated objective) model farm construct for subregional or regional estimates cannot be supported for similar reasons. However, with the appropriate data identified there may be viable alternatives to the currently proposed approach.

Finding 8. A process-based model farm approach that incorporates "mass balance" constraints for some of the emitted substances of concern, in conjunction with estimated emission factors for other substances, may be a useful alternative to the model farm construct defined by EPA (2001a).
Discussion: The mass balance approach, like EPA's model farm approach, starts with defining feeding operations in terms of major stages or activities. However, it focuses on those activities that determine the movement of nutrients and other substances into, through, and out of the system. Experimental data and mathematical modeling are used to simulate the system and the movement of reactants and products through each component of the farm enterprise. In this approach, emissions of elements (such as nitrogen) cannot exceed their flows into the system.

SOURCE: NRC (2002a).

environmental effects of their emissions to water and air have been of increasing concern, especially to those who live nearby. Concerns include potential health impacts of water and air pollution, disagreeable odors, and the possibility of catastrophic events related to extreme weather events. In addition to local and regional effects, animal operations add significantly to the global burden of greenhouse gases, particularly methane and nitrous oxide, which contribute to global climate change (IPCC, 2001). Ammonia in the air contributes to the formation of fine particulate matter over large regions, and its deposition contributes to eutrophication of coastal bays and estuaries.

To understand the historical trends, it is useful to consider the growth that has occurred in the populations of both humans and animals in some locations, made possible by the industrial fixation of nitrogen on a large scale, especially the production of fertilizer by the Haber-Bosch process (the reaction of molecular nitrogen [N_2] with hydrogen [H_2] to make ammonia, which is then used to make other reactive nitrogen [Nr] compounds; see Appendix B). The use of inorganic fertilizers (particularly ammonium nitrate and urea) has greatly increased the production of agricultural crops, especially since 1950, and made possible the rapid increases in the populations of both humans and domesticated animals that have taken place in the twentieth century (Smil, 2001).

Nitrogen flows in U.S. agriculture, including crops and animal production, are shown for base year 1997 in Figure 1-1. (Note that the figure does not include the NO_x generated from the combustion of fuel used to produce and transport crops or to transport animal waste or animal products from farms.) About 18.5 Tg

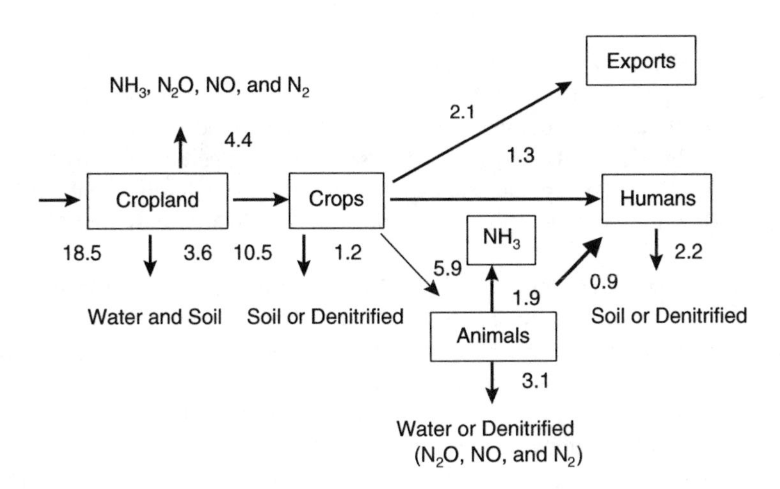

FIGURE 1-1 Mass flows (teragrams of nitrogen per year) of new reactive nitrogen in U.S. agriculture in 1997. Adapted from Howarth et al., 2002.

N/yr (1 Tg = 10^{12} g or 1 million metric tonnes) of new Nr is added to cropland in the form of inorganic fertilizer (11.2 Tg N/yr), cultivation-induced biological nitrogen fixation (5.9 Tg N/yr), and atmospheric deposition of NO_y (1.2 Tg N/yr). Of this, 10.5 Tg N/yr is harvested in crops and 8.0 Tg N/yr is lost from cropland to water and air (some as N_2). Of the 10.5 Tg N/yr in harvested crops, 5.9 Tg N/yr is fed to animals, 1.3 is fed to people, 2.1 is exported, and the remaining 1.2 is lost to the environment (including some N_2). Of the 5.9 Tg N/yr fed to animals, 5.0 Tg N/yr is lost to the environment and the remaining 0.9 is fed to people in the United States. Thus, of the 18.5 Tg N/yr added to cropland from new sources (not including N recycled in manure), humans in the United States consume 2.2 Tg N/yr (all of which is lost to the environment).

In fulfilling their responsibilities to protect human health and the environment, EPA and state and local environmental agencies are grappling with the issues of how to quantify the size and scope of AFO emissions and how to decrease their adverse effects in an economical way based on the best science available. Addressing these adverse effects has implications not only for direct effects on human health and the environment, but also for indirect effects on the structure of the U.S. agricultural economy. The problems caused by animal feeding operations have occurred in part because of the concentration of production in large operations, which is driven by market economics. Actions to decrease the adverse effects (potential economic and social consequences) could lead to the further concentration of production or, alternatively, to a reversal of that trend. Users of this report should be cognizant of these possibilities.

POLICY CONTEXT

Two federal agencies, EPA and USDA, have programs that address the effects of emissions from AFOs. These programs are discussed in detail in Chapter 6. Implementation of programs by both agencies is aided by the states. EPA's programs rely on regulation, while USDA's rely on management, mainly by farmers' actions. The two approaches have the potential to be complementary rather than conflicting if there is close coordination among EPA, USDA, and state governments.

Regulations and programs aimed at mitigating air emissions from AFOs are effective and publicly acceptable only if the information on which they are based is defensible. Public trust can be eroded if regulations and programs appear to be based on poor information. As noted in the committee's interim report, the approach for estimating emissions proposed for use by EPA (2001a) is not adequate because of data limitations and the way in which it characterizes individual farms in relation to emissions (NRC, 2002a). Although current information is not sufficient in many cases to support defensible regulations, EPA is under pressure to take actions to decrease air emissions from AFOs. As a result, the committee believes that EPA must develop new information in a timely manner.

While EPA must rely on quantitative estimates of air emission rates and concentrations of pollutants in the ambient air for regulating air emissions, USDA programs to date have relied less on quantitative estimates of emissions and more on the expected favorable effects of best management practices (BMPs). The committee believes that better understanding of the measurable effects of these practices would help guide USDA in mitigating air emissions and their adverse effects. Thus, both of the federal agencies face much the same issue—a lack of solid scientific information for pursuing their tasks.

SCIENCE CONTEXT

Sound understanding of AFO air emissions and their effects requires the expertise of numerous scientific disciplines, including animal nutrition and metabolism, farm practices, atmospheric chemistry, meteorology, air monitoring, statistics, and epidemiology and toxicology (for health effects). Developing this understanding will also require input from agricultural engineering and economics, and related disciplines. To assemble, integrate, and interpret this broad array of information is a formidable task, especially in view of the different animal types and geographical and climate conditions in which AFOs are found.

EPA has a variety of needs for more accurate estimates of air emissions from AFOs, including the following:

- general monitoring of the nation's air quality;
- determining what pollutants are in the nation's ambient air, their concentrations and their sources;
- identifying the emissions that may have the greatest adverse effects on human health or the environment;
- improving regulatory approaches; and
- assessing the effectiveness of various abatement technologies and strategies.

USDA has a broadly similar need for accurate information, but one that focuses more directly on the kinds of management actions that farmers can take to mitigate emissions at the farm level. Despite numerous reports in the literature, little of the information and analysis now available was designed to meet these needs. There is no comprehensive, sound, science-based set of data on emissions from AFOs. Perhaps the most serious problem in generating such a database is the great variability in AFO emission rates, which cannot be addressed solely by developing more accurate measuring instruments. Emission rates can vary tenfold or more during periods as short as an hour or as long as a year—with changes in the management of the animals, their feed, and weather conditions. Improved accuracy in what is recorded at one location and time may matter very little (Arogo et al., 2001; Mount et al., 2002). This problem can be resolved only by devising a regulatory and management strategy that recognizes and accommodates this variability, combined with a measurement plan that collects sufficient information to address concerns and direct future efforts. This approach will vary with the substance and with the concern. For example, potential effects on human health will generally be inferred from short-term (e.g., hourly or daily) local concentrations (and compositions in the case of particulate matter), whereas effects on climate change will be inferred from regional or national annual emission inventories.

In an ideal world, the data would be accurate and precise, broad in coverage of both substances and AFO types and operations, based on sound sampling plans, timely, detailed (addressing geography, time intervals, climatic conditions, etc.), well documented, in a readily accessible form, and inexpensive. Meeting all of these desirable features will undoubtedly lead to conflicts, so compromises must be made.

CHALLENGES

Facing the need for defensible information on air emissions from AFOs in a timely manner is a major challenge for EPA and USDA. Neither has yet addressed the need for this information in defining high-priority research programs. Neither has asked for nor secured the level of funding required to provide the necessary information. Each has pursued its regulatory and farm management programs under the assumption that the best currently available information can be used to implement its program goals.

The committee believes that the scope and complexity of the information needed by these agencies, as well as the potential environmental impacts of air emissions from AFOs, require a concentrated, focused, and well-funded research effort. Such an effort is described in this report.

STRUCTURE OF THE FINAL REPORT

Chapter 2 describes in broad terms the economics and operating practices of the animal feeding industry and its major sectors (dairy, beef cattle, swine, and poultry). This chapter, along with Chapter 6 on government regulations and programs, sets the stage for the other chapters, which address more directly air emissions and the scientific bases for estimating their rates, concentration, and distribution.

Chapter 3 describes the kinds of air emissions produced by animal feeding operations and their potential impacts on the environment and human health. Chapter 4 examines the state of the science for measuring air emissions, including measurement principles and techniques suited to various on- and off-farm situations. Chapter 5 describes approaches for estimating air emissions from AFOs, including an evaluation of a process-based (mass balance) approach for estimating emissions for each component of the overall production process. In this approach, the origin and ultimate destination of each major chemical element (nitrogen, carbon, sulfur and phosphorus) in the animals' feed is modeled. This chapter also compares the mass balance approach with other approaches for estimating emissions at the individual farm level, within regulatory and management contexts. Chapter 6 describes the legal and programmatic structure for managing—by public oversight and mitigation—air emissions from AFOs.

The information in Chapters 1-6 then serves as the basis for defining both short-term (4-5 years) and long-term (20-30 years) research programs in Chapter 7. The short-term program is designed to provide defensible estimates of air emissions that could be used to support responsible regulation. The long-term program views the animal feeding system more broadly in order to decrease AFO inputs, increase recycling of manure, and eliminate or greatly decrease adverse effects on health and ecosystems. Both the short-term and the long-term research programs are ambitious and will require cooperation and participation by the full range of the research community. Finally, Chapter 8 summarizes the committee's major conclusions and recommendations.

2

Livestock Agriculture and Animal Feeding Operations

INTRODUCTION

Understanding the place of animal feeding operations in the U.S. agricultural economy is a necessary prelude to effective public management of the adverse effects of their air emissions. This chapter starts with information on the overall size of the major livestock feeding operations (cattle, swine, dairy cows, and poultry) and their relationship to crop agriculture. It then turns to the general economics of livestock agriculture and the structure of the livestock industry. It ends with a discussion of the economics of emissions and manure management and potential methods of livestock operation emissions control and mitigation.

LIVESTOCK AGRICULTURE

Livestock agriculture is concerned with raising and maintaining livestock, primarily for the purposes of producing meat, milk, and eggs. Livestock agriculture also includes wool and leather production and may include animals kept for recreation (riding or racing) and draft.

Livestock and livestock products generated from $87.1 billion to $96.5 billion annually (46 to 48 percent of U.S. cash receipts from farm marketings) between 1995 and 1998 (U.S. Department of Commerce, 2000, Table 1109). Livestock agriculture is the market or consumer for a significant portion of U.S. crop agriculture.

- Annual U.S. feed and residual use of feed grains (corn, sorghum, barley and oats) amounted to 154.6 million to 157 million metric tons (1994-

1995 through 1997-1998 crop marketing years), 55 to 63 percent of U.S. feed grain production during this period (USDA, 2000a).

- Corn provided 18.2 percent of cash receipts from farm marketings of crops between 1994 and 1998 (U.S. Department of Commerce, 2000, Table 1109). Sorghum and barley added another 2 percent of cash receipts from farm marketings of crops.
- Hay is consumed by livestock and represented 3.8 percent of cash receipts from farm marketings of crops during this period.
- Livestock agriculture is also the market or consumer for soybean meal and other oilseed meals. Soybeans accounted for 14.7 percent of cash receipts from farm marketings of crops between 1994 and 1998 (U.S. Department of Commerce, 2000, Table 1109).
- Approximately 37 percent of U.S. oilseed output was consumed domestically as oilseed meal during the 1994-1995 through 1997-1998 crop-marketing years (USDA, 1996b, 1997b, 1998, 1999a).

In summary, livestock agriculture directly accounts for nearly half of U.S. cash receipts from farm marketings and provides the market for a significant fraction of the remaining portion of U.S. agricultural output.

The leading states in terms of annual cash receipts from livestock and products in 1997 and 1998, in decreasing order, include Texas ($8.2 billion), California, Nebraska, Iowa, Kansas, North Carolina, Wisconsin, Minnesota, Georgia, Arkansas, Oklahoma, Colorado, and Pennsylvania ($2.85 billion).

In many states, livestock agriculture accounts for more than 65 percent of cash receipts from farming. Examples include Alabama, Colorado, Delaware, New Mexico, New York, Oklahoma, Pennsylvania, Utah, Vermont, West Virginia, and Wyoming (U.S. Department of Commerce, 2000, Table 1113).

Livestock agriculture provides the basis for the meat, dairy, and egg processing industries. Meat products represent 49.8 percent of all non-metro food processing employment and 1 of 16 rural manufacturing jobs (Drabenstott et al., 1999). Finally, meat, dairy products, and eggs are important components of the U.S. diet (Table 2-1).

ECONOMICS OF LIVESTOCK AGRICULTURE

Economic characteristics of livestock agriculture addressed here include markets and prices, production costs, and industry structure.

Markets for Livestock and Products

Prices for livestock and products are determined in competitive markets. With the exception of federal marketing orders for dairy (see Blayney and Manchester, 2001, for a description of U.S. milk marketing programs), markets for livestock

TABLE 2-1 U.S. Per Capita Consumption of Meat, Dairy Products, and Eggs in 2001

Product	Retail weight per year (kilograms)
Broiler chicken	34.7
Beef	30.0
Pork	22.8
Turkey	7.9
Milk and products	264.4[a]
Eggs	252.6[b]

[a]1998, kilograms of milk equivalent on a milk fat basis (USDA, 1999b).
[b]Number rather than kilograms.

SOURCE: USDA (2002c, p.11)

and products are unrestricted. Producers respond to market prices for livestock and their products and to prices of feed ingredients by increasing production following periods of high profit and decreasing production following periods of losses. Biological lags in production response are a fundamental characteristic of livestock agriculture. The gestation or hatching periods of livestock and poultry plus the period from birth to market weight or to milk or egg production impose minimum times in which livestock and poultry farmers can respond to price or profit signals. This period approaches one year for swine and two to three years for cattle. Broiler producers are able to respond within a few months, while egg and turkey producers may require 6 to 18 months to respond. The result of the lagged response is a cycle in production, prices, and profits as producers are constantly adjusting output by expanding or exiting production. Prices and profits in any single year may not be representative of the equilibrium price and profit of a livestock sector due to the length of cycles in prices and profits. Volatility in prices is evident. Feed cost is generally the largest component of total cost and varies directly with ingredient (corn, soybean meal, hay) prices. Recent U.S. Department of Agriculture (USDA) benchmark cost series show feed to be about 60 percent of the cost of broilers, turkeys, table eggs, and pigs. Feed is more than 70 percent of the benchmark cost of weight gain in high plains cattle feeding operations. Volatile prices for feed ingredients and market animals, combined with biological lags in production response, result in extremely volatile profit margins. Extended periods of losses (sometimes severe) and profits are common in the livestock sector.

Confined animal feeding operations have a large share of the nation's livestock and account for an equal or larger share of the products. For example, beef cattle feedlots with more than 1000 head of cattle, which sold an average of

10,983 head in 1997, accounted for 85 percent of the beef cattle sold (EPA, 2001a). The largest size categories for other kinds of livestock operations have similarly large shares of the number of animals and production. Nevertheless, the large number of operations in even the largest size categories keeps any one, or any group, of them from having sufficient market power to affect the prices of their products. Farms in general, including animal feeding operations, are "price takers." Because of the large number of farms of most kinds, each farm faces an elasticity of demand that is "nearly infinite" (Carlton and Perloff, 1989).

Various methods of vertical coordination between meat processing organizations and animal feeding operations (AFOs) are in use (Martinez, 2002). Broiler, turkey, and some swine processors use production contracts. Production contracts are generally defined as contracts between owners of livestock and independent farmers to have the farmers raise the livestock on their farms. Typical production contracts have the livestock owner (frequently, but not necessarily, a processor) provide livestock, feed, medication, and managerial and veterinary support, while farmers provide buildings, labor and management, land, manure management, utilities, repairs, and supplies in exchange for a fee per head or per pound produced. Marketing contracts or agreements are another method of vertical coordination between processors and livestock producers. Marketing contracts or agreements may be defined as contracts to deliver livestock, and establish the base price and price increments for specific attributes (e.g., weight, condition, backfat depth). Marketing contracts are distinguished from production contracts in that farmers retain ownership of the livestock and provide feed and other inputs until the livestock are delivered to the processor. It is the individual farms, whether they sell on "spot" markets or operate under contract, that produce the air emissions that are the subject of this report.

Producers of livestock and poultry compete in an international market. Beef and pork are both imported and exported. Net exports range from 3 percent of pork production to 18 percent of broiler production. Although exports constitute a relatively small fraction of total production, they add significantly to agricultural income. Increased production costs can decrease the international competitiveness of U.S. agricultural production sectors and shift income to foreign producers. A significant cost increase in the U.S. livestock sector could shift production (and emissions) across political boundaries.

Farm Numbers, Inventory, Farm Size, Production, and Productivity

The number of farms in the United States peaked in 1935 at about 6.5 million and has been declining steadily as farm size and productivity rise. There were 1.91 million farms (defined as places selling at least $1,000 of agricultural products in a year) in the United States in 1997 (USDA, 1999c). The fraction of U.S. farms that keep all types of livestock and poultry has also been declining steadily since about 1910.

Increased specialization has accompanied increased productivity. There has been little change in the number of pigs in the United States since 1920. The number of cows being milked peaked at 25 million in 1944 and has since dropped to about 9 million. Milk production per cow increased markedly from 2073 kilograms per year in 1944 to more than 8,000 kiograms per year in 2001 (USDA, 2002c, 2002d). Annual production of livestock and products has risen steadily over the past century, although production cycles are evident in the data. Also evident is a steady increase in livestock productivity (defined here as the quantity of meat, milk, and eggs produced annually from a given inventory of livestock). Productivity gains arise from an increased number of animals born and raised per breeding animal per year, increased growth rates and market weights of animals intended for slaughter, and increased milk or egg production per animal per year. In addition to producing more from a given inventory of animals, livestock farmers have greatly decreased the quantity of feed required to produce a pound of meat, milk, or eggs. Productivity gains have been accomplished through genetic selection, as well as through improvements in diet formulation and processing, housing and environmental controls (e.g., improved buildings, manure removal, and ventilation), veterinary medical care and medications, and management. Havenstein and colleagues (2002) demonstrate that a 2001 strain of broiler chicken fed a current diet requires about one-third the feed and one-third the time to produce a 4.0 pound (lb) live broiler as a 1957 genetic strain chicken fed a diet used in 1957. Since modern broilers are grown to heavier weights, the actual efficiency gains are altered. The modern broiler raised to 5.9 lbs in six weeks requires about 27 percent of the time and 42 percent of the feed per pound of live bird that the 1957 strain required. The 1957 strain required about 103-105 days to produce a 4.0-pound bird. These productivity gains are consistent with those cited by Martinez (2002, Table 3). Note that reduced feed consumption per pound of product results in a proportionally larger reduction in the quantity of excreta on a dry weight basis. For example, if feed consumption is reduced to 42 percent of the original quantity, and if 15 percent of the original quantity was and is retained in the product, then the dry weight excreta would be 31.7 percent of the original quantity excreted ([0.42 – 0.15] / [1.0 – 0.15]).

Farm Size, Production and Market Organization, and Contracts

Dairy. In the United States, there were 79,318 dairy farms with more than three milk cows reported in the 1997 census of agriculture (Kellogg, 2002). Of these, 16 percent were very small (<35 USDA animal units [AUs]), 33 percent were small (35 to 70 USDA AUs), 40 percent were medium sized (70 to 210 USDA AUs) and 9.8 percent were large (>210 USDA AUs). USDA animal units differ from Environmental Protection Agency (EPA) animal units (Appendix E) and are equivalent to 454 kg (1000 pounds) live weight accounting for all animals on the farm. In contrast to other food animal industries, the dairy industry is not vertically integrated. Farms are owned and managed independently of processors. Most dairy farms raise their own replacement heifers but sell bull calves. Fluid milk is sold to processors, which may be controlled by cooperatives or by private or public corporations. (See Blayney, 2002, and Manchester and Blayney, 1997, for further exposition of structure and trends in the U.S. dairy sector.)

Beef Cattle. The number of beef cattle in the United States peaked at 132 million head in 1975. USDA estimated that in 2001 the U.S. cattle inventory was 96.7 million head and that there were 1.05 million cattle operations (operations with at least one or more head of beef or dairy cattle). Many of these are cow-calf operations, with cattle fed on pasture, that are not considered AFOs. For example, 0.65 million cattle operations had fewer than 50 head of cattle and accounted for 11.5 percent of the United States cattle inventory in 2001 (USDA, 2002e). Feedlots vary in size, from a great many operations that hold only a few animals to a small number with a one-time occupancy capacity of more than 100,000 head.

The cattle feeding industry has not developed integration or contractual arrangements to the extent that the poultry or swine industries have. Most feedlots are privately held; an owner may have more than one, but ownership of a feedlot does not necessarily mean ownership of the cattle being fed there. Custom feeding is common where an investor who owns the cattle may have no active involvement in cattle feeding or agriculture except through an investment portfolio.

Cattle farmer-feeder operations are those in which much of the feed used in the feedlot is derived from owned or rented cropland that is part of the operator's overall agricultural operation. These operations may involve feedlots with capacities as large as 10,000-12,000 head. Most farmer-feeder operations probably have a one-time capacity of <2500 head. Large commercial feedlots may have a substantial land base for feed production but in most instances would have to purchase a significant portion of the feed needed.

Custom feeding (housing and feeding cattle on a feedlot for a fee; the cattle are not owned by the feedlot owner) is common. Cow-calf operators who do not have a feedlot may also utilize custom feeding after their cattle have been weaned. The proportion of custom-fed cattle within a feedlot is not necessarily related to overall size of the feedlot. It has become increasingly common for smaller farmer-feeder operations to use custom feeding as a way to decrease risk or to capitalize expansion.

Pigs. Almost all of the U.S. inventory of pigs in each of the three phases of production is housed in buildings. There were 81,130 farms with at least one pig on December 1, 2001. Of those, 84.6 percent had fewer than 1000 pigs in inventory and maintained 13.5 percent of the 58.8 million pigs in the country; 8.6 percent have at least 2000 pigs in inventory and maintained 74.5 percent of the U.S. inventory of pigs (USDA, 2001).

The U.S. markets for pigs include a mix of spot markets, contracts, and processor ownership. For example, USDA (2002a) indicates that 14.1 percent of market hog sales on October 21, 2002, were spot market transactions (where prices are negotiated within 24 hours of the delivery of pigs to market); another 67.4 percent were conducted through marketing contracts. The remaining 18.5 percent of hogs slaughtered that day were packer owned. USDA also estimates that 33 percent of the U.S. pig inventory on December 1, 2001, was under production contract to operations that owned at least 5000 pigs (USDA, 2001). Many of the entities that own pigs and contract them out under production contracts are pig producers and not pork processors. Some pork processors own pig farms, and some own pigs and contract them out to farmers under production contracts. Some Midwestern states including Iowa prohibit packer ownership of pigs prior to slaughter.

Poultry. Almost all broilers (young chickens raised for meat) and turkeys are raised in buildings, as are egg-laying chickens. Martinez (2002) indicates that more than 80 percent of broilers are raised under production contracts and the remainder are raised on farms owned by the processors. He also reports that 56 percent of turkeys are raised under production contracts and another 32 percent are owned and raised by turkey processors. Martinez (2002) indicates that 60 percent of chicken eggs are produced on farms owned by the processor and another 38 percent are produced under production contracts for the processor.

Although not substantially concentrated economically in terms of being able to affect prices for their output, the animal feeding operations (as distinguished from the large processing firms, referred to as "integrators" in the case of swine and poultry) are regionally concentrated (Box 2-1). The cumulative shares of production based on number of animals for the top four and next four states are shown in Table 2-2.

As improvements have been made in poultry housing, and equipment for feeding, watering, and ventilation, the number of birds that an individual farmer could care for has increased. A flock of 1000-2000 birds was considered huge in the 1920s. Presently, one broiler farmer can easily manage and care for 150,000 or more birds. Complexes housing laying hens for the production table eggs may have 1.5 million birds that are typically managed by a crew of approximately 15. Again, economics have caused poultry farmers to look for more efficient and effective methods of producing more animals per unit of labor.

State and regional specialization, as shown in Table 2-2, is the result of various factors that affect the livestock industry. The cost of animal feed, the importance of which is evident in the frequent high rankings of states in the Midwest, is obviously significant. Transportation costs—both for getting feedstuffs to the feeding operations and for getting products to markets—are also important, although their importance tends to be reduced by practices such as shipping feed grains in unitized trains, which can significantly lower transportation costs for

BOX 2-1 Poultry Production in the United States

Poultry production in the United States was essentially a farm sideline until the 1930s. Economically disadvantaged farmers, primarily in areas of the country where soils, climate or other conditions were not conducive to traditional row crop agriculture, were the early pioneers in transforming poultry production into a primary farming opportunity. For example, poultry production is purported to have begun in north Georgia due to the continued failure of cotton crops in the region. Farmers were desperate to develop an alternative. In northwest Arkansas an apple blight was the economic incentive, and on the Delmarva peninsula, declines in shellfish harvests and disease problems in the region's traditional truck farming (fruit and vegetable) crops made farmers desire a reliable cash crop (Gordy, 1974).

As the industry evolved, it looked for ways to become more efficient. Jesse Jewell in Gainesville, Georgia is generally credited with advancing the idea of vertical integration in poultry production. He understood that bringing hatcheries, feed mills, and processing plants together as coordinated units would greatly improve scheduling and reduce costs. Vertical integration resulted in an infrastructure being developed (hatchery, feed mill, processing plant) that further localized poultry production into regions. It was advantageous from a transportation standpoint for all of these aspects of poultry production to be in close proximity. Generally the farmers who produced the poultry were located within 50 miles of the feed mill. Thus, the concentration of the poultry industry in discrete areas of the United States has been due to economics (Sawyer, 1971).

large operations. Other factors such as climate, differences in cost of labor and land, population density, and state regulation of the livestock industry are also important, but their effects are not obvious in the rankings.

TABLE 2-2 Leading Livestock Production States by Animal Sector

Sector	Top Four States	Percent	Next Four States	Percent	Total
Beef cattle	TX, KS, NE, CO	60	IA, CA, OK, MN	16	76
Milk cows	WI, CA, NY, PA	44	MN, TX, MI, ID	17	61
Swine	IA, NC, MN, IL	57	IN, NE, MO, OH	21	78
Broilers	GA, AR, MS, NC	48	TX, VA, DE, MO	16	64

SOURCE: EPA (2001a).

Animal Feeding Operations and Concentrated Animal Feeding Operations

Animal feeding operations and concentrated animal feeding operations (CAFOs) are classifications of livestock and poultry farms used by the U.S. Environmental Protection Agency for regulation under the Clean Water Act. An AFO is defined as "a lot or facility where animals have been, are, or will be stabled and confined and fed or maintained for a total of 45 days or more in any 12 month period, and where crops, vegetation forage growth, or post harvest residues are not sustained in the normal growing season over any portion of the lot or facility" (40 CFR part 122.23(b)(1)). An AFO is defined as a CAFO if it confines more than 1000 EPA animal units at any time during the year. EPA defines animal units differently than USDA. One EPA animal unit of each type of livestock is indicated in Table 2-3.

Threshold farm sizes are published by EPA to distinguish CAFOs for each animal type. An AFO is defined as a CAFO (based on EPA regulations prior to December 15, 2002) if it has more than the following numbers of animals of any species: 1000 feeder and slaughter cattle, 700 mature dairy cattle, 2500 swine weighing more than 55 pounds, 55,000 turkeys, 100,000 laying hens or broilers if the facility has continuous overflow watering or 30,000 laying hens or broilers if the facility has a liquid manure system, and 5000 ducks. The critical distinction between AFOs and CAFOs is that CAFOs are potentially regulated as point sources and required to obtain National Pollutant Discharge Elimination System (NPDES) permits. These definitions are solely for the purpose of determining which farms are to be regulated by various methods, with the largest farms receiving the most stringent oversight.

USDA has a different definition of an animal unit, which can lead to confusion in comparing EPA and USDA statistics, including confusion in estimating air emissions because of differences in the animal base on which estimates of air emissions are predicated. The committee suggests that estimates of air emissions in the future be based on a modeling approach that is more flexible than has been

TABLE 2-3 Number of Animals per EPA Animal Unit

Animal Type	Head
Slaughter or feeder cattle	1.0
Mature dairy cow	0.7
Pigs weighing 25 kg or more	2.5
Turkeys	55.0
Chickens	100.0
Sheep or lambs	10.0
Horses	0.5

used to date and that is consistent with continuous, rather than periodic, estimates of animal growth. This leads to Finding 1:

FINDING 1. Much confusion exists about the use of the term "animal unit" because EPA and USDA define animal unit differently.

RECOMMENDATION: Both EPA and USDA should agree to define animal in terms of animal live weight rather than an arbitrary definition of animal unit.

Production Systems

Production systems vary substantially across the country and from farm to farm. This section describes basic elements of the most prevalent systems.

Dairy. Most dairy farms are diversified crop and animal production systems. Some feeds are purchased, but dairy producers usually grow their own forages (whole plant feeds such as hay or silage) and raise their own replacement stock. Most dairy farmers sell their bull calves and raise heifers as replacement animals. The advantage of raising heifers on farm is that it helps prevent introduction of diseases when animals are introduced to the milking herd. In a typical herd, mature cows calve every 12 to 14 months, producing a female calf 50 percent of the time. Milk production per day increases for about 10 weeks and then decreases for the remainder of lactation. Typically, the lactation period lasts about 10 to 12 months. Some farmers use bovine somatotropin injections in mid-lactation to sustain higher amounts of milk production per day. Cows are bred artificially when behavioral and physiological signs of ovulation occur about 60 to 120 days after calving. Lactation continues until two months prior to the next predicted calving. Cows are culled from the herd and slaughtered for low-grade meat production because of failure to become pregnant, low milk production, or chronic health issues. Calves, growing heifers, and dry cows are often housed separately from lactating cows. Young calves are frequently housed in separate hutches or grouped together with animals of similar age in pens or pasture. Replacement heifers are bred, usually by artificial insemination, between 14 and 17 months of age and calve 9 months later. A typical herd with 100 lactating cows may also include 18 dry cows and 86 growing heifers (Dunlap et al., 2000) for a total inventory of 204 head. Young dairy calves consume casein or soy-based milk replacer until adjusted to grain and eventually forage-based diets as they mature. Lactating cattle in peak production consume diets with as much as 60 percent of dry material from grains and high-energy by-products and 40 percent from forages (whole plant crops such as hay or silage). Lactating cattle at lower levels of production and mature cattle between lactations consume diets comprised mostly of forages.

Beef Cattle. Most of the cattle in feedlots in the United States are referred to as yearlings. They enter the feedlot weighing 340 to 410 kg and are fed high grain diets for 130 to 150 days. They are harvested at an average of about 590 kg. There are wide variations on this theme, so these generalizations are less accurate than for other specie production systems. As an example, Holstein steer calves are commonly placed in feedlots on high grain diets when they weigh 160 kg and fed for more than 300 days. In backgrounding yards, calves enter the feedlot weighing 180 to 230 kg and are usually fed high-roughage diets until they weigh 360 to 410 kg. These cattle may then be sold to another feedlot for finishing, or they may remain in the same feedlot and be fed high-grain diets.

Large feedlots typically have a continual movement of feeder cattle in and finished cattle out. Occupancy will have seasonal highs and lows, but there are always cattle on feed. Many smaller operations feed one group of cattle each year. In these systems, calves (500 pounds or 227 kg) enter the feedlot in the fall and are marketed the following summer. During a portion of each year, these operations have no cattle on feed. Many combinations of these production system themes exist in the industry.

Feedlot designs vary by region and type of operation. The most common design is an open pen with 0-15 percent of the surface paved. The balance of the pen surface is earthen. Space allocations range from 70 to 500 square feet per animal. The proportion of paving applied to the pen surface increases in regions that receive more rainfall. Typically, area-per-animal allotments decline as more paving is used.

Bedding is not generally used in earthen pens with large area allocations per animal. Bedding during winter months (and in some instances year-round) is used in paved pens. It is common to include housing in colder or higher-precipitation regions. When housing is provided with open pens, the housing is generally paved. Shedded area allocation is approximately 20 square feet per animal, and bedding is used only in winter months. Feed bunks are usually included in the housed area of these operations.

Total-confinement systems refer to pens completely under roof. Some systems use partial or fully slatted floors with either deep (storage) pits or shallow pits that are flushed or scraped. Other systems have paved floors and use bedding throughout the year. Space allocations will be as low as 25 square feet per animal in total-slat, deep-pit facilities and 40 to 50 square feet in paved floor, bedded, confinement barns.

Pigs. Almost all pigs are raised in total confinement. Pig farms are organized around three phases of production. Farrowing operations maintain a breeding herd of mature females and produce weaned pigs that are typically 3 or 4 weeks old and weigh 5.4 to 7.3 kilograms. Nursery operations receive the weaned pigs and produce feeder pigs that are typically 10 to 11 weeks old and weigh 20-27 kilograms. Finishing operations receive feeder pigs and feed them to market weights of 113 to 127 kilograms at about 25 weeks of age. Various combinations of these production phases may be found on a single site. Farrow-to-finish operations include all three phases. Farrow-to-feeder pig operations include farrowing and nursery phases. Wean-to-finish operations include nursery and finishing phases.

Finishing pigs are usually allotted 7 to 8 square feet of space and housed in pens with constant access to feed and water. Nursery pigs also have constant access to feed and water, are housed in pens, but have less space. Sows kept for farrowing have more space and may be fed individually two or more times daily to maintain health. Recent finishing buildings are designed to house 800 to 1220 growing pigs each. An individual finishing farm may have two to six or more finishing buildings. Nursery buildings may have several rooms and house 2500 weaned pigs. One full-time person can provide the routine daily labor required by 4000 to 5000 nursery pigs or growing hogs. Sow farms consist of facilities for gestation and breeding as well as for farrowing. About one-twentieth of the sow herd is bred, farrows, or weans pigs each week. This rotation reflects the pigs' gestation cycle and provides a constant flow of pigs from the operation. Large, specialized farrowing operations may house 1200 sows or multiples thereof and employ one person for every 200 or 250 sows in inventory. Such operations may average more than nine pigs weaned per litter and 2.2 or more litters per year per sow in inventory. Annual production exceeds 20 pigs weaned per sow per year.

Pig buildings may be ventilated naturally with ridge vents and fabric curtain sides that can be opened. Other pig buildings are ventilated by fans mounted in the walls. Tunnel ventilation is used in warm climates to cool poultry and livestock by pulling a large volume of air in one end of the building and out the other end with large fans. Much lower rates of ventilation are used in cooler seasons and in cooler climates. Pig feed consists primarily of ground corn, soybean meal, and supplemental minerals and vitamins. Feed is often ground, mixed, and pelleted at large centralized feed mills, although some farms still grind their own corn and mix in soybean meal and vitamin-mineral premixes. Diets are tailored to the nutrient requirements of the pigs at various stages of growth and reproduction (e.g., NRC, 1998a). Whole-herd feed conversion rates have fallen steadily and are now well below 3 pounds of feed per pound of live pig produced in some production systems.

Poultry. Broilers and turkeys are raised in similar systems. A centralized feed mill produces pelleted diets consisting of ground corn, soybean meal, and mineral and vitamin supplements. Specialized farms maintain breeding flocks and produce hatching eggs. Hatching eggs are collected at a hatchery where chicks and/or poults are hatched, separated by gender, and delivered to farms for grow-out to market weight. Depending on the market being served, some broiler flocks are now marketed at 6 weeks of age or less. Others are raised to much heavier weights at 7 to 8 weeks of age for further processing or for sale as roasters.

Turkey hens are generally marketed as whole birds at 12 to 14 weeks (5.4 to 7.2 kilograms) or for further processing at 18 to 20 weeks (8.1 to 10.0 kilograms). Toms are generally marketed at 35 to 40 pounds at 20 to 22 weeks of age, and almost all toms are now processed further.

Broiler houses will handle 20,000 to 30,000 birds per house, and farms generally have two to six such houses. Turkey buildings generally hold 6500 to 8500 toms or 13,000 to 17,000 females. Tunnel ventilation is used in warm climates, while open-sided buildings with lower rates of ventilation are generally used in cooler seasons and climates. Some turkey farms have both brooding and growing facilities (generally with one brooder for two growing facilities), but most, due to disease-related problems in multiaged operations, are now moving to all-in, all-out operations. Turkeys and broilers (as well as nursery pigs and finishing pigs) are generally raised on an all-in, all-out basis. That is, a flock of day-old birds is placed in an empty building and raised to market weight. The house is then emptied and cleaned prior to the arrival of the next flock a week or two later. Most broiler farms are organized in "complexes" consisting of a centralized feed mill, hatchery, and centralized broiler processing plant, with grow-out farms located within a 50-mile radius of the plant. Turkey complexes are similar, although turkeys are generally transported far greater distances.

Most table eggs are produced in buildings with the hens in cages. There were 72,616 farms with at least one pullet or laying hen older than 13 weeks in the United States in 1997 (USDA, 1999c, Table 21). These farms housed 367 million pullets and hens. There were 606 farms with at least 100,000 pullets and hens 13 weeks or older that housed 65 percent of the U.S. flock. Feed is primarily ground corn or other grain and soybean meal with vitamin and mineral supplements. Almost all egg production facilities are enclosed and are power ventilated.

Manure Management

Manure management varies widely across species, region, and farm type. Since manure management can have a significant effect on emissions, attention is given here to some of the common systems.

Manure management systems are generally based on USDA recommendations from the Natural Resource Conservation Service (NRCS), Cooperative Extension Service (MWPS, 1992). Manure management systems vary with climate, soil productivity, farm size, and other factors. The systems in use now reflect research, development, education, and regulatory programs over the past 40 years. For example, Humenik (2001) provides a history of the evolution of anaerobic lagoon and sprayfield systems corresponding to the development of the Clean Water Act in 1972.

Dairy. There are many different systems for handling dairy manure. Tie-stall barns (cattle confined in stalls) often have gutters that can be cleaned by mechanical scrapers. Most U.S. dairies with fewer than 100 cows use this means for cleaning barns (USDA, 1996a). Free-stall barns are often cleaned using mechanical scrapers that pass through the alleyway. Most farms with more than 200 cows use this means of cleaning (USDA, 1996a). Flush systems are increasingly common on large farms. However, flush systems require greater storage capacity than mechanical scrapers because more liquid is added to the animal manure despite recycling from a storage pond or lagoon. Dry lots or bedded packs can be used to house cattle in dry climates, with manure removed only occasionally with a tractor. Dairy cattle manure is either stored dry in piles on concrete or earthen pads, stored as a slurry in a concrete or lined lagoon or storage tank, or mixed with flush water in earthen or lined lagoons which may be covered with biological material (e.g., straw), covered with impermeable material (e.g., synthetic polymers), or left uncovered.

Beef Cattle. Manure management in feedlots varies with the range of facilities described previously. Earthen-floor pens are routinely scraped, and the solids are collected into mounds within the pens. The manure mounds are removed on schedules that depend on the climate, region, and class of cattle involved. Solids removal from these systems may occur monthly, quarterly, semiannually, or annually. Some feedlots do not remove the manure yearly; rather a mound is created in the fall and peeled over winter, allowing the manure to dry in summer and be mounded again. The one-turn-per-year feedlots typically remove solids only once a year. When there is a continuous flow of cattle and pens are on feed less than 150 days, solids removal likely coincides with the sale of cattle from a pen.

Pens with extensive paving require regular (weekly, semiweekly) removal of solids. Primary factors affecting the frequency of scraping are stocking density in the pen, precipitation, and use of bedding. Solid-floor, total-confinement barns with bedding are generally cleaned every month.

In all of these systems, the disposition of removed solids depends on season and region. It is often necessary to stockpile solids at a location outside the pen until the material is spread onto cropland, perhaps weeks or months later. Some operations compost the solids, but this practice is not prevalent because of climatic conditions, costs, and additional management requirements.

Permitted feedlots with outside pens have runoff controls ranging from vegetative filters to settling basin pond systems to lagoons. Settling basins are handled as solid waste usually when the material is dry. Ponds may be allowed to evaporate or be used as a source of irrigation water. Lagoons are pumped, usually each spring and fall, with liquid manure applied to cropland.

Slatted-floor confinement designs with flush systems typically incorporate some degree of solids separation to allow recycling of flush water. The high solids content effluent fraction would be stored in lagoons or slurry store-type structures. Deep-pit facilities are usually emptied each spring and fall.

Local ordinances are having an increasing influence on manure handling and management. These are highly variable and often specific to an individual feedlot. The result of federal, state, and local regulations and stipulations is a checkerboard of manure management strategies. This creates confusion in the permitting process, may accommodate specific optimums by location, and may lead to a real or perceived disparity of requirements.

Pigs. Manure management for pigs varies widely with climate, geographical characteristics, and size and type of operation. A small proportion of farms in Iowa and other states has adopted a deep-bedded system in the past decade, in which pigs are kept in hoop buildings on deep straw beds. The bedding material and manure are removed periodically and spread on land. More prevalent systems include slurry handling systems, common in the upper Midwest, and anaerobic lagoon and flushing systems, with land application of liquid lagoon effluent, common in the Southeast. A variant of the anaerobic lagoon system can be found in the arid West where liquid is evaporated rather than applied to cropland. The slurry handling systems include collection of manure, spilled water and feed, and wash water in under-floor concrete pits or gutters. The floor of the pig buildings consists partially or totally of concrete gang slats, steel tribar, or woven wire such that manure can fall through gaps in the flooring. The undiluted manure is referred to as slurry and may contain 5 to 10 percent solids. The slurry may be stored in a deep pit beneath the building, or it may be pumped to an outside storage tank (usually open topped and made of concrete or glass-lined steel) or an earthen slurry basin. Slurry is pumped out of storage and applied to land with tractor-drawn equipment in either the fall or the spring. The application rate is limited to the amount of manure that will meet the plant available nitrogen requirements of the crop to be produced there. A recently revised NRCS standard has caused some producers to shift to applying manure to more land, at a lower rate that will not exceed the plant available phosphorus requirements of the crop.

The anaerobic lagoon and sprayfield system of manure handling is characterized by an anaerobic treatment and storage lagoon with a flushing or pit recharging system for frequent removal of manure from the buildings. Concrete slats or other flooring with openings allow manure, spilled water, and feed to fall into a shallow pit or a flush gutter beneath the floor. In the pit recharge system, less than 2 ft of liquid depth is maintained in the shallow pit and a standpipe-plug is pulled on a regular schedule to allow the liquid and accumulated manure to drain to the anaerobic lagoon. The pit is then recharged with lagoon liquid. The flush system does not maintain liquid in the flush gutter, but a flush tank at the higher end of the building is filled with several hundred gallons of lagoon liquid and released into the flush gutter every few hours. The flush liquid and accumulated manure drain into the anaerobic lagoon. The anaerobic lagoon is a large earthen structure in which a minimum treatment depth of several feet of liquid must be maintained at all times. This treatment depth maintains an anaerobic environment that supports anaerobic microbes employed to digest the organic matter in the manure. In addition to the treatment volume, the lagoon is also designed to contain temporary storage volume (six months to one year of manure volume and rainfall accumulation), emergency storage (a 25-year, 24-hour storm accumulation, plus a chronic rainfall accumulation in some states), sludge accumulation depth, and freeboard. Lagoon effluent generally has less than 1 percent solids and a small fraction of the nutrient content of manure slurry. Liquid lagoon effluent is land-applied using automated irrigation equipment. Liquid effluent is applied at a rate that meets the plant available nitrogen or phosphorus requirements of the crop. Annual land application volume is equal to the volume of manure, spilled water and feed, water used to wash the building interior, and rainfall accumulated in open structures, minus evaporation from barns and open structures. A variant of the anaerobic lagoon system uses the high rate of evaporation and low rainfall in some locations to decrease effluent volume.

Broilers and Turkeys. Many broiler and turkey grow-out buildings have earthen floors. The floor is covered with a bedding material such as wood shavings to collect and dry the manure. The relatively low moisture content of poultry manure makes this approach practical. The bedding material and accumulated manure (called litter) are generally removed from the buildings and replaced once each year. The surface of the litter is generally raked to remove feathers and caked material, and then new shavings are added between flocks. Once removed, the litter is generally directly land-applied, but it may be stacked and stored in covered piles or in a litter storage shed until it is loaded into a manure spreader (a truck- or tractor-drawn implement) and land-applied. In arid regions, thin bed drying may be used.

Eggs. A variety of manure management systems are used for layer operations. Most caged layer buildings have concrete floors. In the high-rise layer system, manure falls onto a concrete floor, accumulates there, and is removed periodically as a dry material that can be spread mechanically on land. Anaerobic lagoon and flushing systems have also been used on layer farms, but are becoming less and less common. There are also cage systems with manure belts that pass beneath the cages and convey the manure to a collection point. The manure is then augured out of the building for storage until it is eventually spread on land.

Economics of Emissions and Manure Management

Farmers generally behave as profit maximizers; that is, they try to use inputs and produce products such that the difference between total revenue and costs is maximized. Farm practices to limit emissions and manage manure can be considered in this context. Since manure management can affect rates and composition of emissions, it is given considerable attention in this and the following section.

Farmers are willing to incur costs to store, transport, and land-apply manure up to the value of additional revenues generated and costs avoided. In the case of manure management, the costs avoided include the purchase and application of commercial fertilizer. Costs avoided may also include those associated with nuisance complaints. In some cases, manure utilization is thought to increase yields more than commercial fertilizer. Such a yield increase would be an example of additional revenue generated. An example of the economic definition of a waste product would be if the costs of utilizing manure as a fertilizer exceed the value of benefits generated. A product that costs more to use than the value of benefits generated by its use is a waste. Once a product is identified as a waste, profit-maximizing behavior seeks the least cost (total cost minus total revenue) option for waste disposal.

Manure treatment (as opposed to simple storage and land application) may become the most profitable or least costly option in some circumstances (e.g., Drynan et al., 1981). A variety of factors affect the economic attractiveness of treatment. High transportation and land application costs, low commercial fertilizer prices, and low treatment costs create incentives for manure treatment. High transportation costs arise from long distances between livestock and fields. Hauling distances are increased by having small and noncontiguous fields, low-yielding soils and crops (low fertilizer requirement per acre), higher nutrient concentrations in manure, larger farm sizes, and by regulations. Some costs of treatment decline (on a dollar-per-gallon basis) as farm size increases. Manure treatment may include stabilization (decomposition of organic matter to prevent odor and flies), decreased pathogens, concentration of components that must be transported (such as nutrients), separation of low-value material (e.g., water, organic matter, grit) for application to nearby land, or other modification of form to produce more useful by-products.

Emissions and manure management become a policy issue when not all costs and benefits of livestock production are realized by the farmer. Costs and benefits realized by others in the absence of a negotiated exchange (purchase or sale) are referred to as externalities. Negative externalities are costs incurred by others, such as loss of environmental quality or adverse health effects. Positive externalities are benefits received by others such as increased income, employment, and improved public services arising from a larger tax base.

Policy is generally designed to maximize social welfare by maximizing total benefits (private and public) minus total costs (private and public). Where externalities are present, governments may adopt policy to intervene in the market. Intervention may take the form of regulation and enforcement, investment in research and education, and/or support for the development of markets that allow externalities to be partially internalized. A maximizing social welfare solution may be difficult to identify; it is more feasible to identify policy changes that increase social welfare. A policy change that creates benefits that are valued more than the costs imposed is one that increases social welfare. Thus, the policy objective with respect to emissions and manure management on livestock farms may be to limit emissions to the rate at which the value of marginal benefit (marginal health and environmental damage avoided) is greater than or equal to the marginal cost (marginal cost of emissions mitigation to farmers, the community, and consumers). Critical components of the benefits estimation procedure include (1) accurate measurement of the marginal changes in emissions due to various mitigation strategies, (2) accurate measurement and prediction of changes in environmental quality and public health that arise from such changes in emissions, and (3) accurate estimation of the dollar value that society places on the marginal changes in environmental quality and public health. Critical components of the cost estimation procedure include (1) accurate measurement of the incremental investment, annual operating costs, occasional costs, and operating revenue incurred by the farmer to adopt each mitigation strategy; (2) estimation of the distribution across farms of farmers' responses to the additional cost (continuing to operate, altering or decreasing production, or closing the operation); (3) estimation of the effect on equilibrium production and prices across regions, states, and countries; and (4) estimation of the secondary loss of income, employment, and property tax base in communities that lose livestock production. (See Chapter 5 of the committee's interim report for further exposition of cost-benefit analysis; NRC, 2002a).

Efficient policy change can be defined as a change in policy such that no other policy would generate the same value of benefits at lower cost or generate greater benefits at the same cost.

A final important consideration in policy change is the Pareto criterion. This criterion requires that no one be made worse off by a policy change and at least one person be made better off. If a policy truly creates benefits of greater value than the costs imposed, then those receiving benefits can compensate those bearing the costs and still be better off than they were. The costs of a policy change to individual farmers and to communities may be inadvertently overlooked in a national comparative statistical analysis comparing the equilibria before and after a policy change. The costs of transition can be great where policy change has different effects across regions. Application of the Pareto criterion decreases the displacement during a transition by compensating those bearing the costs. Elimination or minimization of individual welfare loss decreases opposition to policy change.

Where manure is considered a waste or a product of little value, farm practices to limit emissions and to manage manure are driven by regulatory requirements (such as the EPA CAFO rule and state rules) and nonregulatory guidelines (such as NRCS standards and Cooperative Extension Service recommendations).

Costs and benefits of manure utilization have not been well documented in surveys, but some budget estimates (with their inherent limitations) are available. Regulatory requirements and nonregulatory guidelines are important to cost analyses of various manure management systems if they affect the rate at which manure can be land-applied and the size of storage and treatment structures. Drynan et al. (1981) have published a detailed analysis of manure management costs for several systems applied to North Carolina swine farms. They concluded that "the cost estimates suggest that almost all operations will choose to use a lagoon in preference to hauling manure with tank wagons." Cox (1993) budgeted costs of irrigation systems for lagoon effluent on various sizes and types of pig farms in North Carolina. His estimates of total irrigation cost per 1,000 gallons of effluent were in the range of $1.50 to $2.00. Lorimor and collegues' (1999) survey of custom rates for tractor-drawn drag-line injection of manure slurry in Iowa reflected rates averaging $6 to $8 per 1000 gallons injected, with additional charges if the slurry had to be transported more than 1 mile. Roka (1993) budgeted total costs and value of fertilizer saved for lagoon and sprayfield systems in North Carolina and slurry systems in Iowa. Zering (1998, 1999) adapted the budgets provided by these authors and calculated finishing farm costs of $1.10 to $1.90 per hog finished for an anaerobic lagoon and sprayfield system in North Carolina. He also calculated costs of $2.85 per hog finished in an Iowa finishing operation and value of fertilizer saved at $2.44 for a net cost of $0.42 per hog finished. Note that the lagoon system was the least costly alternative in North Carolina, while the slurry system was less expensive in Iowa. These results are consistent with the differences in field size, crop yield, and climate (anaerobic lagoons must be up to 40 percent larger in cooler climates to achieve the same level of treatment) between the two states and the observed practices. Each of these estimates is a result of a series of assumed coefficients; together they illustrate the sensitivity of resulting estimates to changes in each parameter and variable.

Information needs arising from the economics of emissions and manure management are substantial. Several critical components of cost and benefit estimation are listed earlier in this section. Accurate measurement of emissions from current and proposed livestock production and manure management systems is one of the most critical components. The economic basis for measurement of emissions is that society cannot rationally decide how much cost to incur to decrease emissions without knowing the extent to which emissions will be decreased and the value of the benefits that will be generated by that decrease.

Alternative Manure Management and Emission Mitigation Strategies

Air emissions from livestock and poultry farms arise from many sources spread across the entire farm and the emissions are matters of concern. Sources include manure storage and handling facilities within and outside buildings, transport and land application of manure and effluent, and feed storage and handling facilities. Options for control or mitigation of air emissions from livestock and poultry operations are limited. Several research efforts around the country involve some of the technologies and management practices that may prove useful in decreasing air emissions from AFOs. Some technologies not discussed here may prove as efficacious as those listed. Discussion of possible emission modification or control strategies is presented in broad categories including strategies for animal feeding, animal health, and manure management.

Animal Feeding and Animal Health Strategies. Animal feeding strategies to protect the environment have been studied closely in recent years (e.g., Kornegay, 1996). A possible method to decrease emissions is to decrease the source of the material being emitted. Several approaches for decreasing the quantity of nitrogen excreted in manure are available. One approach is to continue to increase the productivity of livestock and poultry. Increasing production per animal (faster growth rate, increased milk production) decreases the number of animals required to fill the market demand for those products. The animal's requirements can be divided into needs for maintenance (maintaining basal metabolism) and production. Meeting maintenance requirements results in a fixed amount of nitrogen excretion for each animal in the herd or flock. Since fewer animals are required with increasing production, the nitrogen losses to manure are decreased. Dunlap et al. (2000) showed that increasing milk production of dairy cows—by administering bovine somatotropin, increasing photoperiod using artificial lighting, and milking three times daily instead of two—would decrease manure nitrogen by 16 percent for a given amount of milk produced. Increased productivity has been accomplished through genetic selection, improved diet, improved housing and environmental controls, improved veterinary medical care, and improved management. Animal health is important to emissions control since unhealthy animals have decreased growth or decreased milk or egg production but their maintenance needs to remain the same, and they continue to produce emissions and manure.

A second approach to decreasing the quantity of nitrogen excreted is to more precisely match diets to requirements of groups of animals at various stages of growth, reproduction, lactation, and egg production. Since most animals are fed in groups, diets are composed to meet or exceed the requirements of all or nearly all of the animals within the group. Like human beings, animals also have species-specific requirements for essential amino acids (NRC, 1994, 1998a, 2000, 2001a). Grouping animals with similar requirements enables meeting the requirements of each animal more closely with the same diet. For example, grouping growing animals by age and gender allows a substantial decrease in the amounts of nutrients fed and excreted. Feeding broilers four different diets during their grow-out period, rather than the standard practice of three diets, resulted in decreasing nutrient requirements by 5 percent (Angel, 2000). (This practice is referred to as phase feeding.) Grouping dairy cows into separate production groups on a farm was predicted to decrease nitrogen excretion by 6 percent compared to feeding all lactating cows the same diet (St-Pierre and Thraen, 1999). Such reductions have great economic importance since profit margins tend to be small.

Many commercial operations have already adopted phase feeding; all-in, all-out production; and separate gender feeding.

A third approach is to increase the precision with which digestible or metabolizable amino acid, mineral, energy, and other nutrients in the diet match the current requirements of the animal. Feeding amino acid supplements has had the greatest impact of all recently adopted practices on decreasing nitrogen excretion to manure. Animals require a specific profile of amino acids for optimal production, which most feeds do not provide. When balancing the diets of animals, corn products and legumes are typically mixed to provide a complementary set of amino acids. Corn is high in methionine but low in lysine, while legumes are the reverse. By blending grain and soybean meal diets to ensure adequate inclusion of the most limiting amino acids, nutritionists invariably include excess quantities of other amino acids (included in crude protein). Synthetic amino acid supplements can be used to further decrease protein feeding without sacrificing production or health. Sutton et al. (1996) showed that corn and soybean meal diets for growing pigs supplemented with lysine, tryptophan, threonine, and methionine decreased ammonia and total nitrogen in freshly excreted manure by 28 percent. Amino acids protected from degradation in the rumen of cattle have been developed and shown to decrease needs for feed nitrogen by approximately 10 percent (Dinn et al., 1998). Exclusion of feed ingredients that are not highly digestible or metabolizable by animals decreases the quantity excreted. Some researchers are also examining the inclusion of enzymes and other compounds to increase the digestibility of feed ingredients. Feed efficiency is expected to continue improving for the foreseeable future.

Increased precision in diet formulation may preclude the feeding of some crop and food processing by-products because their digestibility is low or their nutrient composition profile does not match that required by the animal. As a result, this material may become waste and be land-applied or otherwise disposed of.

Rapid changes in feed efficiency and resulting excretion rates leave many published coefficients obsolete. There is a need for recurring measurement of typical performance and updating of published numbers for variables such as volume excreted per day, nitrogen excreted per day, volatile solids, and biochemical oxygen demand (BOD) or chemical oxygen demand (COD) of excretion per day. This need for updated data is apparent in attempts to budget nitrogen emission factors using dated estimates of nitrogen excretion.

Manure Management Strategies. A wide variety of manure management technologies and strategies have been considered over the last 30 years (e.g., ASAE, 1971). The systems and strategies now in wide use are those that proved the most cost-effective and reliable at achieving their design objectives. For the most part, those objectives did not include minimization of emissions of ammonia or methane, but rather focused on odor and dust control, avoidance of direct discharge to surface water, and land application at agronomic rates. Recent attention to air emissions reveals that very few data exist on emissions of some compounds from these systems. This section is intended to highlight air emission issues related to some of the manure management technologies being considered. It is important to keep in mind that water quality protection, nuisance avoidance, animal environment protection, and worker health protection remain as considerations in manure management system design, not to mention cost and risk minimization.

Manure naturally undergoes microbial decomposition that produces a number of inorganic gases and organic compounds. Manure handling and treatment can have a large influence on the physical, chemical, and biological properties of manure and consequently on the production and emission of gaseous compounds. Many treatment technologies are available that may be important in emission mitigation. However, the effectiveness of most of these technologies is not well quantified. Some technologies may decrease emissions of certain gases or compounds but increase those of others (e.g., sprinkling water on feedlots to suppress dust emissions may increase organic decomposition and the emission of ammonia and odorants). Other technologies may suppress emissions during one stage of manure management only to increase those in subsequent stages. A complete farm system approach to emissions measurement is required. Treatment technologies have to be analyzed with clear objectives as to what emissions are to be mitigated. Two white papers recently published by the National Center of Animal Manure and Waste Management (Lorimor et al., 2001; Moore et al., 2001) review various animal manure handling and treatment technologies that have been used on farms, or have been extensively researched.

Recently, USDA NRCS initiated a project to identify and evaluate the emerging animal manure treatment technologies that will most likely be used by animal producers in the next 5 to 10 years. The following discussion includes manure handling and treatment technologies that have been identified by the project and have relevance to air emissions (Melvin, personal communication, 2002).

Storage covers for slurry storage tanks, anaerobic lagoons, and earthen slurry pits are being studied as a method to decrease emissions from these containments. Covers being studied, both permeable and nonpermeable, range from inexpensive chopped straw (on slurry containments only) to more expensive materials such as high-density polyethylene. Covers can decrease emissions from storage, but their net effect on emissions from the system depends on how the effluent is used on the farm.

Anaerobic digestion in closed containment has been studied for many types of applications. This is the process that occurs in anaerobic lagoons. When conducted in closed vessels, gaseous emissions including methane, carbon dioxide, and small amounts of other gases (possibly ammonia, hydrogen sulfide, and volatile organic compounds) are captured and can be burned for electricity generation or water heating, or simply flared. An in-ground digester being tested on a swine farm in North Carolina is an example of the ambient temperature version of this technology (there are also mesophilic and thermophilic designs). The concentration of ammonia remaining in effluent from that digester may be higher than that which can be volatilized from lagoon effluent once exposed to air. The primary effect of anaerobic digestion is to decrease the amount of volatile solids (corresponding to COD or BOD). Pathogens may also be decreased in the process. Complete anaerobic digestion substantially decreases odor. Emissions from combustion of digester gas should be measured. The State of California recently awarded a $5 million grant to Inland Empire Utilities Agency to develop a centralized waste processing facility in Chino, California, and also provided $10 million as cost sharing for dairy farmers to build anaerobic digesters.

Aeration of liquid or solid waste streams is accomplished by mechanically forcing air through the waste. The objective of aeration is to maintain some concentration of (dissolved) oxygen in the waste stream to support aerobic microbes that digest the organic material in the manure. Aerobic digestion generally produces carbon dioxide rather than methane, and decreases the amount of ammonia produced, producing nitrate and organic forms of nitrogen instead. Aerobic treatment is generally more expensive than anaerobic treatment because of the equipment, electricity, repairs, and management required. Westerman and Zhang (1995) found that the typical electricity cost to completely treat finishing hogs' manure using aeration was $14 per pig space per year at $0.09 per kilowatt-hour (kWh). This amounts to $5.38 per hog finished at 2.6 groups per year. These authors found that $2.34 in electricity costs per hog finished would be required to attain partial odor control with aeration. Aerobic treatment produces several times the volume of sludge produced by anaerobic digestion. Costs, benefits, and emissions arising from sludge management must be considered.

Solid-liquid separation is used on some farms now and is being considered as part of several alternative manure management systems. Zhang and Westerman (1995) reviewed engineering studies of solids separation. They reported that from none to roughly half of total Kjeldahl nitrogen (TKN), phosphorus, and COD can be removed. Costs of separation increase as the fraction separated increases and as the use of polymers increases. The costs, benefits, and emissions from solids storage and land application after separation are important considerations. Solids separation may decrease the volatile solids load on a subsequent treatment process and may increase the land required to receive swine manure nutrients. Further treatment (composting or dewatering) may add to cost but allow less expensive transport off the site. Effects of solid separation on odor concentrations at the property line remain to be determined.

Composting is a method of stabilizing organic solids and decreasing pathogens by allowing aerobic or anaerobic microbes to digest the material. Composting requires space, labor, and management and can affect emissions positively or negatively. Its primary benefit is to decrease volume and produce a more acceptable soil amendment.

Other manure management technologies and strategies have been or are being considered (e.g., Burton, 1997; Miner et al., 2000) and an extensive research program is under way. Many of these have been applied in municipal and industrial settings (e.g., Crites and Tchobanogolous, 1998). On July 25, 2000, Smithfield Foods, Inc., entered into a voluntary agreement with the Attorney General of the State of North Carolina to provide resources to be used in an effort to develop innovative technologies for the treatment and management of swine wastes that are determined to be technically, operationally, and economically feasible (Williams, 2001). Performance standards require comprehensive analyses of odor and ammonia emissions, pathogens, and economics for each technology. Currently, 18 technologies or systems are being studied.

Other technologies and practices such as livestock housing design and operation affect air emissions. A considerable research and development effort has been devoted to evaluation of inexpensive filters for exhaust air from buildings and of "windbreak walls" to deflect and disperse the exhaust airstream from buildings. Land application methods to decrease emissions are also being studied.

In summary, many options of varying cost and effectiveness are being evaluated for reducing emissions and managing manure on livestock and poultry farms. Measurement of air emissions from existing and alternative systems on commercial farms is needed for both emissions of local concern and those of regional and national concern.

SUMMARY

The structure and management practices of the animal feeding sector respond mainly to economic dictates as influenced by government regulations. Both economic factors and regulations affecting this sector change as understanding of their effects and the effects of responses to them also change. This chapter provides what amounts to a recent snapshot of the sector's structure and operations. While the exact direction of changes in economic factors and regulations, and thus the future structure and operation of the industry, may not be predictable, users of this report should expect change to occur.

3

Air Emissions

INTRODUCTION

This chapter describes the emissions of concern to the committee, their character as possible pollutants, factors that influence their production and dispersion in the atmosphere, their spatial and temporal dynamics, and their potential impacts. This description of the emissions of concern, along with the preceding section that described the livestock industry that produces them, sets the stage for the remaining chapters.

The focus of this report is on air emissions from animal feeding operations (AFOs). This section outlines some aspects of that issue. First, it discusses the chemical species of interest, with special attention to the contributions of AFOs to the inventories of these species (Table 3-1). Next, the various factors that determine the rates of emissions from individual AFOs are discussed. Following this, the dispersion of these emissions is discussed to provide a framework for understanding the relationship between emissions and atmospheric concentrations. Finally, the committee discusses in somewhat greater detail the potential human health and environmental impacts of air emissions of the key species. It should be noted that the material in this section is not meant to be a comprehensive survey, but rather is designed to highlight certain key issues in terms of air emissions from AFOs. The goal is to set the stage for the more detailed analysis of the scientific basis for estimating air emissions from AFOs in Chapters 4 and 5.

SPECIES OF INTEREST

Ammonia

The nitrogen in animal manure can be converted to ammonia (NH_3) by a combination of hydrolysis, mineralization, and volatilization (e.g., Oenema et al.,

TABLE 3-1 Annual Anthropogenic Emissions of Constituents of Concern, 1990[a]

	NH_3 (Tg N)		N_2O (Tg N)		NO[b] (Tg N)		CH_4 (Tg C)		VOCs (Tg mass)	
	Global	U.S.	Global	U.S.	Global	U.S.	Global	U.S.	Global	U.S.
Energy										
Fossil fuel combustion + production	0.1	0.0	0.2	0.1	21.2	6.7	70.1	16.3	67.3	10.2
Biofuel combustion	2.2	0.2	0.1	0.0	1.3	0.1	11.5	0.3	31.6	0.9
Industrial Processes	0.2	0.0	0.5	0.1	1.5	0.1	0.6	0.0	55.7	12
Agriculture										
Agriculture and natural land	12.6	1.0	1.0	0.1	5.2	0.4	44.4	0.4	NA[c]	NA
Animals	21.1	1.4	1.0	0.1	1.0	0.1	68.7	5.7	NA	NA
Biomass burning										
Savannah burning	1.8	0.0	0.1	0.0	3.1	0.0	4.8	0.0	4.9	0.0
Deforestation	1.4	0.0	0.0	0.0	1.1	0.0	4.2	0.0	7.8	0.0
Waste										
Agriculture waste burning	1.3	0.1	0.1	0.0	2.2	0.2	8.9	0.7	13.8	1.2
Landfills	2.7	0.1	0.0	0.0	0.0	0.0	26.5	7.5	0.0	0.0
Total	43	3	3	0.4	37	8	240	31	181	24
Percent from animals	48	50	33	25	1	1	19	18	NA	NA

[a] H_2S emissions are not available for the level of disaggregation shown for other species, but they are small relative to other S sources (e.g., SO_2 from fossil fuel combustion) on a national and global basis. They might be important on a regional basis in some areas.
[b] Estimates of NO emissions from manure applied to fields vary substantially. Reported values for the fraction of manure nitrogen lost as NO have been as high as 5.4 percent, but 2 percent was chosen as a midrange value. Uncertainty is approximately a factor of two.
[c] VOC emissions are not available for agricultural sources except agricultural waste burning.

SOURCE: van Aardenne et al. (2001).

2001). Urea in the urine of mammals can be hydrolyzed rapidly to ammonia and carbon dioxide by urease enzymes present in feces. On a global scale, animal farming systems emit to the atmosphere ~20 Tg N/yr as NH_3 (Galloway and Cowling, 2002), an amount that comprises about 50 percent of total NH_3 emissions from terrestrial systems (van Aardenne et al., 2001). Based on gridded emissions documented by van Aardenne et al. (2001), about 3 Tg N/yr was emitted from natural and anthropogenic sources in the United States in the mid-1990s. Emissions from animal waste (1.4 Tg N/yr) accounted for about 50 percent of the total. This figure is similar to the 1.9 Tg N/yr in Figure 1-1 (Howarth et al., 2002) estimated to come from animal emissions.

Once emitted, the NH_3 can be converted rapidly to ammonium (NH_4^+) aerosol by reactions with acidic species (e.g., HNO_3 [nitric acid] and H_2SO_4 [sulfuric acid]) found in ambient aerosols. Gaseous NH_3 is removed primarily by dry deposition, while aerosol NH_4^+ is removed primarily by wet deposition. As an aerosol, NH_4^+ contributes directly to PM2.5 (particulate matter having an aerodynamic equivalent diameter of 2.5 μm or less) and, once removed, contributes to ecosystem fertilization, acidification, and eutrophication. After NH_3 is emitted to the atmosphere, each nitrogen atom can participate in a sequence of effects, known as the nitrogen cascade (see discussion of environmental impacts and Figure 3-2 later in this chapter) in which a molecule of NH_3 can, in sequence, impact atmospheric visibility, soil acidity, forest productivity, terrestrial ecosystem biodiversity, stream acidity, and coastal productivity (Galloway and Cowling, 2002). Excess deposition of reactive nitrogen can also decrease the biodiversity of terrestrial ecosystems (NRC, 1997). Since the residence times of NH_3 and NH_4^+ in the atmosphere are on the order of days, a regional-scale perspective is necessary to assess the environmental effects of, and control strategies for, NH_3 emissions.

Nitrous Oxide

Nitrous oxide (N_2O) forms and is emitted to the atmosphere via the microbial processes of nitrification and denitrification. Global emissions in 1990 were ~15 Tg N/yr (Olivier et al., 1998), of which anthropogenic sources accounted for ~3 Tg N/yr. Of these, N_2O emissions from animal excreta accounted for ~1 Tg N/yr (Olivier et al., 1998; van Aardenne et al., 2001). In the United States, total anthropogenic sources in 1990 were ~0.4 Tg N/yr, with animal excreta contributing about 25 percent (Table 3-1) (van Aardenne et al., 2001).

N_2O diffuses from the troposphere to the stratosphere, where it is lost to photolysis and other processes. Once emitted, N_2O is globally distributed because of its long residence time (~100 years); it contributes to both tropospheric warming and stratospheric ozone depletion. N_2O has a global warming potential 296 times that of carbon dioxide (CO_2) (IPCC, 2001).

Nitric Oxide

Anthropogenic activities, especially combustion of fossil fuels, account for most of the nitric oxide (NO) released into the atmosphere (van Aardenne et al., 2001). Nitrification in aerobic soils appears to be the dominant agricultural pathway to NO. Direct emissions of NO from livestock and manure are believed to be relatively minor, but a substantial fraction of manure nitrogen applied to soils as fertilizer can be emitted as NO. However, emissions from agricultural systems are discussed briefly as a whole because of the direct link between livestock agriculture and feed-grain agriculture in the United States.

The contribution of soil emissions to the global oxidized nitrogen budget is on the order of 10 percent (Finlayson-Pitts and Pitts, 2000; Seinfeld and Pandis, 1998; Stedman and Schetter, 1983). Where corn is grown extensively, the contribution is much greater, especially in summer; Williams et al. (1992a) estimated that contributions from soils amount to about 26 percent of the emissions from industrial and commercial processes in Illinois, and may dominate emissions in Iowa, Kansas, Minnesota, Nebraska, and South Dakota. The fraction of fertilizer nitrogen released as NO depends on the amount and form of nitrogen (reduced or oxidized) applied to soils, the vegetative cover, temperature, soil moisture, and agricultural practices such as tillage. A small fraction of NH_4^+ and other reduced nitrogen compounds in animal manure can also be converted to NO by microbial action in soils.

Nitric oxide and nitrogen dioxide (NO_2) are rapidly interconverted in the atmosphere and are referred to jointly as NO_x. In turn, NO_x can be incorporated into organic compounds such as peroxyacetyl nitrate (PAN) or further oxidized to HNO_3. Gas-phase HNO_3 can be converted to aerosol nitrate (NO_3^-) (e.g., by reaction with ammonia). The sum of all oxidized nitrogen species (except N_2O) in the atmosphere is often referred to as NO_y. The residence time of NO_y is of the order of days in the lower atmosphere, with the principal removal mechanism involving wet and dry deposition of HNO_3 and aerosol NO_3^-. In terms of environmental effects, NO_x is an important (and often limiting) precursor in tropospheric ozone (O_3) production. Furthermore, NO_3^- aerosol is a contributor to PM2.5, and nitrogen deposition in the forms of HNO_3, and aerosol NO_3^- can have ecological consequences as mentioned earlier.

Methane

Methane (CH_4) is produced by microbial degradation of organic matter under anaerobic conditions. Total global anthropogenic CH_4 is estimated to be 320 Tg CH_4/yr (corresponding to 240 Tg C/yr [teragrams of carbon per year]) (van Aardenne et al., 2001), comparable to the total from natural sources (Olivier et al., 2002). Of the various anthropogenic sources, the agricultural sector is the largest, with livestock production being a major component within this sector (van Aardenne et al., 2001). In the United States, livestock emissions contribute 7.6 Tg CH_4/yr (5.7 Tg C/yr) of a total anthropogenic source of 41 Tg CH_4/yr (31 Tg C/yr) (van Aardenne et al., 2001).

The primary source of CH_4 in livestock production is ruminant animals. Ruminants (sheep, goats, camels, cattle, and buffalo) have unique, four-chambered stomachs. In one chamber, called the rumen, bacteria break down grasses and other feedstuff and generate CH_4 as one of several by-products. The production rate of CH_4 is affected by energy intake, which is in turn affected by several factors such as quantity and quality of feed, animal body weight, and age, and varies among animal species and among individuals of the same species (Leng, 1993). CH_4 is also emitted during anaerobic microbial decomposition of manure (DOE, 2000). The most important factor affecting the amount produced is how the manure is managed, because some types of storage and treatment systems promote an oxygen-depleted (anaerobic) environment. Metabolic processes of methanogens lead to CH_4 production at all stages of manure handling. Liquid systems tend to encourage anaerobic conditions and to produce significant quantities of CH_4, while more aerobic solid waste management approaches may produce little or none. Higher temperatures and moist conditions also promote CH_4 production.

Methane is destroyed in the atmosphere by reaction with the hydroxyl (•OH) radical. Because of its long residence time (~8.4 years), CH_4 becomes distributed globally. Methane is a greenhouse gas and contributes to global warming (NRC, 1992); it has a global warming potential 23 times that of CO_2 (IPCC, 2001).

Volatile Organic Compounds

Volatile organic compounds (VOCs) vaporize easily at room temperature and include fatty acids, nitrogen heterocycles, sulfides, amines, alcohols, aliphatic aldehydes, ethers, *p*-cresol, mercaptans, hydrocarbons, and halocarbons. Total emissions of VOCs from all sources in the United States were estimated to be 30.4 Tg/yr in 1970 and 22.3 Tg/yr in 1995 (EPA, 1995a).

The major constituents of AFO VOC emissions that have been identified include organic sulfides, disulfides, C_4 to C_7 aldehydes, trimethylamine, C_4 amines, quinoline, dimethylpyrazine, and C_3 to C_6 organic acids, along with lesser amounts of aromatic compounds and C_4 to C_7 alcohols, ketones, and aliphatic hydrocarbons.

Hydrogen Sulfide

Hydrogen sulfide (H_2S) is produced in anaerobic environments from the microbial reduction of sulfate in water and the decomposition of sulfur-containing organic matter in manure. On a global basis, 0.4-5.6 Tg S/yr of reduced sulfur gases (mostly H_2S and dimethyl sulfide) are emitted from land and sea biota (Penner et al., 2001). Most atmospheric H_2S is oxidized to sulfur dioxide (SO_2), which is then either dry deposited or oxidized to aerosol sulfate and removed primarily by wet deposition. The residence time of H_2S and its reaction products is of the order of days.

The short lifetime of H_2S, coupled with the fact that H_2S emissions from AFOs on a global (Schnoor et al., 2002) and national (NAPAP, 1990) basis are small relative to other atmospheric sulfur sources (e.g., soils, volcanoes, wetlands, fossil fuel combustion), means that H_2S emissions from AFOs contribute relatively little to ecosystem effects. However, it appears that H_2S emissions on a regional basis could be important contributors to the sulfur burden of the atmosphere for those regions with a high density of AFOs and few other sources. Emission inventories of H_2S for these regions are necessary to explore this issue further. H_2S may also have local effects of concern, especially odor.

Particulate Matter

In this report, the committee considers particulate matter as PM10 and PM2.5. PM10 is commonly defined as airborne particles with aerodynamic equivalent diameters (AEDs) less than 10 μm. The number refers to the 50 percent cut diameter in a Federal Reference Method PM10 sampler where particles of 10 μm AED are collected at 50 percent efficiency (62 Fed. Reg. 38651-38701). Similarly, PM2.5 refers to the particles that are collected in a Federal Reference Method PM2.5 sampler, which has a 50 percent cut diameter of 2.5 μm (62 Fed. Reg. 38651-38701). The classification is important from the perspective of this report because of the manner in which PM is regulated (see Chapter 6).

AFOs can contribute directly to primary PM through several mechanisms, including animal activity, animal housing fans, and air entrainment of mineral and organic material from soil, manure, and water droplets generated by high-pressure liquid sprays, and they can contribute indirectly to secondary PM by emissions of NH_3, NO, and H_2S, which are converted to aerosols through reactions in the atmosphere. Particles produced by gas-to-particle conversion generally are small and fall into the PM2.5 size range. Key variables affecting the emissions of PM10 from feedlots include the amount of mechanical and animal activity on the soil-manure surface, the moisture content of the surface, and the fraction of the surface material in the 0-10 μm size range.

The AED of PM is critical to its health and radiative effects. PM2.5 can reach and be deposited in the smallest airways (alveoli) in the lungs, whereas larger particles tend to be deposited in the upper airways of the respiratory tract (NRC, 2002b). Smaller particles are also most effective in attenuating visible radiation, causing regional haze.

Odor

Odor from AFOs is not caused by a single species but is rather the result of a large number of contributing compounds (including NH_3, VOCs, and H_2S), Schiffman et al. (2001) identified 331 odor-causing compounds in swine manure. A further complication is that odor involves a subjective human response. Though research is under way to relate olfactory response to individual odorous gases, odor measurement using human panels appears to be the method of choice now and for some time to come. Since odor can be caused by hundreds of compounds and is subjective in human response, estimates of national or global odor inventories are not currently possible. Odor is a common source of complaints from people living near AFOs, and it is for local impacts that odor has to be quantified.

Two methods are reported in the literature for how to define odor intensity. A standard developed and used in Europe (European Committee for Standardization, 2002) defines odor unit (OU) as the mass of a mixture of odorants in 1 m^3 of air at the odor detection threshold (ODT)—the concentration of a mixture that can be detected by 50 percent of a panel. The standard further defines the odor concentration of a sample as the number of OUs in 1 m^3 of sample, which is numerically equivalent to the dilution ratio required to dilute the sample to the ODT. Others define OU as a unitless odor concentration, which is numerically equivalent to the numerical factor by which an air sample must be diluted until the odor reaches the ODT.

Other Substances

During the course of this study, the committee was informed through its scientific sessions and public forums of other potential substances (e.g., bioaerosols, pesticides, and carbon disulfide emitted through the air from AFOs) that should be considered for this final report. The committee was also informed by sponsors of their priority for the committee to report on those substances (NH_3, H_2S, N_2O, VOCs, PM, and odor) listed in the Statement of Task (Appendix A). In the interim report, the committee reached a consensus to add NO to this list.

As the committee deliberated its final report, consideration was given to the availability of information on other substances, the Statement of Task, sponsors' priorities, and the time available to meet sponsors' needs. Although the committee might have liked to explore these other substances, it did not do so because of those considerations. Likely, much less scientific information exists on measurement protocols and the importance of AFOs in the emission of other substances; however a lack of discussion in this report should not be construed as an indication of a lack of their potential importance. A brief discussion has been included on bioaerosols (Appendix C).

FACTORS AFFECTING AIR EMISSIONS

Climatic and Geographic Differences

Differences in climate influence emissions from AFOs because of differences in temperature, rainfall frequency and intensity, wind speed, topography, and soils. Increases in mean ambient temperature and moisture are expected to increase gaseous emission rates from several components of AFOs, including manure storage and land to which manure has been applied. Averaging reported emission factors does not remove the effects of these climate factors, especially if the emission factors selected were determined mostly in one climatic region of the country. This bias remains when emission factors for one type of animal are applied to others by adjustments to reflect differences in excretion rates.

Differences in emissions from AFOs may also arise because of other geographic differences, such as availability of land for manure or lagoon effluent utilization, rates of evapotranspiration, and differences in soil texture and drainage that can impact application rates of lagoon water, or differences in soil microenvironments that affect microbial action and the resulting gaseous emissions. The breed of a given animal species (e.g., selected for cold or heat tolerance) and feed formulations (because of changes in animal maintenance requirements) may also vary in response to geographic and climatic differences.

It is difficult to project how these various sources of uncertainty will combine to influence gaseous emissions and whether these factors will have significant impacts on percentages of nitrogen, carbon, or sulfur lost in gaseous species averaged over a year's time. Climatic differences do not negate the mass balance flows of elements through AFOs (as discussed in Chapter 5), so that, unless there is a significant change in storage of an element within the manure management system, changes in total emissions (to air and water) can come about only because of changes in excretion resulting from changes in feed formulation or efficiency of animal nutrient utilization. Differences may not be as important for annual emissions of major gaseous species (e.g., NH_3, CH_4) as for shorter-term emissions of PM and odors.

Hourly, Daily, and Seasonal Changes

Changes in emissions from individual AFOs due to hourly, daily, and seasonal variations are discussed here because measurements to characterize emissions have usually been conducted for short periods of time. Failure to account for short-term cycles in an experimental design could result in significant systematic errors in a derived annual emission factor. Table 3-2 illustrates seasonal variations in ammonia emission fluxes (mass/area-time) from primary anaerobic swine lagoons. Within one study (Harper et al., 2000), there is as much as a 12.5-fold variation in measured ammonia flux during one summer season. While some of

TABLE 3-2 Measured Emission Fluxes of Ammonia from Primary Anaerobic Swine Lagoons as a Function of Measurement Method and Period

Measurement Method[a]	Period	TAN[b] (mg/L)	Emission Flux[c] (kg NH_3-N/ha-d)	Reference
Micromet.	August-October	917-935	73-156[d]	Zahn et al. (2001)
Micromet.	Summer	230-238	3.2-40	Harper et al. (2000)
Micromet.	Winter	239-269	1.3-1.9	Harper et al. (2000)
Micromet.	Spring	278-298	3.1-9.8	Harper et al. (2000)
Micromet.	August	574	15.4-22	Harper and Sharpe (1998)
Micromet.	January	538	4.7-12.1	Harper and Sharpe (1998)
Micromet.	May	741	5.2-15.4	Harper and Sharpe (1998)
Micromet.	Summer	193	2.9-8.4	Harper and Sharpe (1998)
Micromet.	Winter	183	6.0-9.1	Harper and Sharpe (1998)
Micromet.	Spring	227	3.0-6.6	Harper and Sharpe (1998)
Chamber	August	587-695	34-123	Aneja et al. (2000)
Chamber	December	599-715	5.3-28	Aneja et al. (2000)
Chamber	February	580-727	1.3-10	Aneja et al. (2000)
Chamber	May	540-720	12.3-52	Aneja et al. (2000)
TG OP-FTIR	May	—	37-122	Todd et al. (2001)
TG OP-FTIR	November	—	7.8-67.6	Todd et al. (2001)
Chamber	September	101-110	0.57-3.5	Aneja et al. (2001)
Chamber	November	350	0.46-1.73	Aneja et al. (2001)
Chamber	February-March	543-560	0.72-5.39	Aneja et al. (2001)
Chamber	March	709-909	0.82-2.95	Aneja et al. (2001)
Chamber	April-July	978-1143	104	Heber et al. (2001)
Chamber	May-July	326-387	39	Heber et al. (2001)

[a]Micromet. = micrometeorological; Chamber = dynamic flow-through chamber; TG OP-FTIR = tracer gas open path Fourier transform infrared spectroscopy.
[b]TAN = total ammoniacal nitrogen concentration (milligrams per liter), except for the Aneja et al. (2000) entries, which are for TKN (total Kjeldahl nitrogen) because TAN was not available. TAN is typically 45-95% of TKN.
[c]Emission fluxes are in kilograms of nitrogen as NH_3 per hectare per day.
[d]Half of the lagoon was covered and half was uncovered; the lower numbers were measured from the covered half.

SOURCE: Data are from Arogo et al. (2001, Tables 9 and 10). Lagoon surface areas varied from 0.39 to 3.07 ha.

the variability in fluxes in the table is probably due to variability in temperatures and pH, there is little information in the scientific literature to suggest what other factors are important.

AFOs are essentially collections of different biological systems—each operating with its own hourly, daily, and seasonal cycles. At the scale of the individual animal, there are daily cycles in the activities of eating, defecating, and moving about (the latter is particularly important for generating PM from cattle feedlots). Some microbial cycles that produce emissions may be closely tied to animal activity through the amount and frequency of defecation. As an animal grows, the amount and composition of its feed intake change, as do the amount and composition of its manure (NRC, 1994, 1998a, 2000, 2001a). This gives rise to corresponding changes in microbial activity and emissions. Lactating animals experience changes in productivity throughout their natural cycle, with changes in feed consumed and nutrients excreted (NRC, 1998a, 2000, 2001a). Although the capacity of an AFO may remain essentially constant, different numbers of animals may occupy this space during the year, depending on the production cycle used. Thus, the cycling of animals through an AFO is another source of variation in emissions.

Upsets in the daily rhythms of animals may result in changes in feed ingested and nutrients excreted, and may last for a period of several days. Such upsets may occur due to illness, drastic short-term changes in weather, or breakdowns of farm equipment. Depending on the manure management system employed, such event-driven changes may not have significant effects in terms of emissions of NH_3 or CH_4 but may have a major impact on other emitted species such as VOCs and PM. Other event-driven processes that can increase emissions include lagoon turnover, flush cycles for housing units, manure scraping at feedlots, and land application of lagoon liquids (EPA, 2001a).

The impact of daily cycles and upsets on estimates of annual emissions may not be important, provided a sufficient number of observations are made to account for them. However, given the current paucity of emissions data for the development of emission factors for AFOs, it is not possible to determine to what extent such cycles and upsets may have affected published emission measurements. As discussed in the interim report, (NRC, 2002a, Chapter 2), averaging does not compensate for the systematic bias that may be present as a result of the failure of an experimental design to account adequately for such events.

Animal Life Stage

Reference has already been made to changes in feed formulations that occur during the life cycles of most animals produced at AFOs and their subsequent effects on the amount and composition of fecal matter and urine excreted (NRC, 1994, 1998a, 2000, 2001a). Figure 3-1 (NRC, 2002a) provides an example of changes in the rate of nitrogen excreted for "grow-finish" swine produced at AFOs in the southeastern United States. The data are based on a growth model (ARC, 1981) used by a commercial swine producer to adjust feed formulations. Data have been normalized to 100 percent for the highest rate of nitrogen excretion to prevent the disclosure of proprietary information. Additional examples of nitrogen excretion across species in varying stages of growth and production are provided in Appendix D.

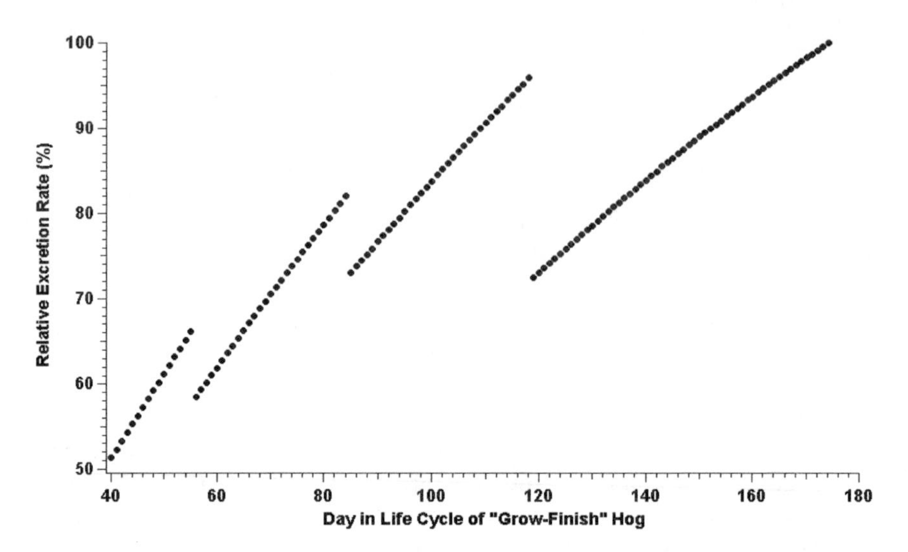

FIGURE 3-1 Relative excretion rate of nitrogen versus day in the life cycle of a grow-finish hog at a commercial swine production facility in the southeastern United States. Animals attain the designation of grow-finish hog at approximately day 40 in their life cycle and are finished at about day 174.

As expected, the relative amount of nitrogen excreted daily tends to increase as the pig grows, reflecting changes in the daily total nitrogen consumed. The actual feed formulation is changed four times during the growth cycle of the hog to account for changes in nitrogen required for maintenance and growth. Sharp decreases in the relative amount of nitrogen excreted per day when the formulation changes are not simply an artifact of the model, but reflect periods of adjustment by the animal to the changes in feed composition. Overall, there is a series of curvilinear increases in the amount of nitrogen excreted per day for finishing swine under this model, with nitrogen excretion nearly doubling during the latter half of the animal's growth period. The emphasis in Figure 3-1 is on total nitrogen excreted. Expressed as a percentage of body weight, the nitrogen excreted would actually decrease throughout the life cycle.

Figure 3-1 illustrates that if daily housing emissions of NH_3 are directly related to daily nitrogen excretion, and the model is an accurate representation of nitrogen excretion, there will not be a simple increase in emissions from the confinement unit with time. Thus, averaging together emission measurements made from several different housing units with different age animals, or from the same housing unit during different times in one growth cycle, may significantly under- or overestimate emissions, depending on the age of the animals when samples are taken. Actual emissions, however, will also depend on the manure collection practices (flush frequency, pit recharge, pull plug, or pit storage) associated with the confinement unit. A manure collection practice that accumulates manure for relatively long periods of time, such as pit storage, may act to smooth the variations in nitrogen emissions due to variations in daily excretion. At a minimum, the data displayed in Figure 3-1 demonstrate that the same sampling scheme may not be applicable to all swine confinement units and that measurements of emissions may have to be weighted to account for differences in animal age.

Management

Optimal management is vital to the success of individual AFOs for the production of quality animals and should also result in decreased emissions. For example, appropriate drainage and manure removal minimize PM generation from cattle feedlots (Sweeten et al., 1998). Effects of animal health on feeding habits are important to maintain consistent nutrient uptake efficiency and prevent feed spoilage. This attention includes maintenance of proper ventilation for animals in confined housing units, maintenance of drainage systems to remove manure on a frequent basis, and regular (perhaps daily) visual inspection of animals and their daily routines. Adherence to nutrient management plans will decrease the potential for excessive air emissions or surface runoff that results from overapplication of nutrients to crops. Anaerobic lagoons should not exceed design-loading rates and should be maintained in the proper pH range for waste stabilization.

Quantitatively assessing the overall impact of effective management on decreasing emissions is currently not possible due to the paucity of emissions data. However, management practices should be included in assessing emissions from individual AFOs. A summary of milk production for 372 dairy farms in the Mid-Atlantic region of the United States (Jonker et al., 2002; Table 3-3) demonstrates the effects that management practices can have on nitrogen utilization efficiency (NUE). Each management practice has its own distinct effect on milk production and NUE. Feeding a total mixed ration increased milk production by 6.6 percent with no significant change in NUE. Use of bovine somatotropin resulted in a significant increase in both milk production and NUE, thereby decreasing the amount of nitrogen excreted per unit of production. While farms that practiced seasonal calving had significantly lower milk production, NUE was unchanged.

Roles of Microorganisms in Emission Rates

Microbes play important roles in the generation of gases emitted at various stages in animal production systems. Microbial activity is primarily responsible for degradation of feed, generation of methane by ruminants, and conversion of animal waste to produce pollutant gas emissions from housing, storage, and land to which manure is applied.

TABLE 3-3 Relationship of Management Practices on 4 Percent Fat Corrected Milk and Nitrogen Utilization Efficiency

Management Practice	4% Fat Corrected Milk (% change)[a]	Nitrogen Utilization Efficiency (% change)[a]
Total mixed ration	6.6[b]	–1.4
DHIA member[c]	10.7[b]	4.4[d]
BST[e]	14.1[b]	6.9[b]
3× milking[f]	9.9[b]	5.3
Extended photoperiod	11.3[b]	4.3[d]
Seasonal calving	–15.1[b]	–1.4
Cover crops	0.4	2.1
Nitrogen nutrient management plan	–0.7	0.0
MUN testing[g]	6.9[b]	2.5
Complete feed[h]	–6.3[b]	–5.6[b]

[a]Percent change from using versus not using the management practice.
[b]$P < .01$. Probability that the use of the technology is actually different than not using the technology.
[c]Dairy Herd Improvement Association membership with routine monitoring for milk production.
[d]$P < .05$. Probability that the use of the technology is actually different than not using the technology.
[e]Bovine somatotropin.
[f]Milking cows three times compared to twice daily.
[g]Milk urea nitrogen (MUN) can be used to assess diet adequacy for lactating dairy cattle (Jonker et al., 1999).
[h]Complete feeds are grain mixtures manufactured for use on multiple farms compared to custom grain mixes used on an individual farm.

SOURCE: Data derived from Jonker et al. (2002, Table 8).

Ruminant livestock, such as cattle, have a digestive system that allows them to eat coarse plant material that humans and other animals cannot digest. This makes it possible to obtain food from land that is not suitable for crop production by having livestock harvest forage and convert it to milk and/or meat. The unique digestive system of a ruminant animal consists of a four-part stomach, which includes the rumen, reticulum, omasum, and abomasum. The rumen is the first and largest compartment, making up about 80 percent of the total stomach volume, and is unique to ruminant animals. In it, microbial organisms such as bacteria, protozoa, and fungi break down and ferment plant material into products that the animal can use for energy.

Methanogenic bacteria, located mainly in the rumen, are responsible for the methane produced in the animal's digestive tract as fibrous materials in feed are broken down. Since methane production results from this action, diets that are high in forages (relative to concentrates) will produce more methane. The use of ionophores in feed (dairy cattle are excluded) causes a temporary decrease in methane production by suppressing methanogenic bacteria in the rumen. Although this effect is temporary (as bacteria become resistant to the effects of the ionophore), ionophores do serve to increase the efficiency of feed utilization.

DISPERSION OF AIR EMISSIONS—METEOROLOGICAL CONSIDERATIONS

Temporal Dynamics

An atmospheric substance can be characterized by its lifetime (also called its residence time) in the atmosphere—defined as the time required (in the absence of sources) to decrease its concentration to $1/e$ (where e is the base of the system of natural logarithms; $1/e$ is approximately 0.37) of the initial concentration. The chemical species of interest in air emissions from AFOs span a wide range of lifetimes. Soluble species (i.e., NH_3, some VOCs) have lifetimes equivalent to those of water in the atmosphere, about 1 to 10 days, depending on precipitation. Reactive species such as NO and H_2S have lifetimes on the order of days or less before they are oxidized to other more water-soluble species such as nitric and sulfuric acids. The lifetimes of VOCs are controlled by the rates of hydroxyl radical attack and water solubility, and range from hours to months. Methane has a much longer lifetime of about 8.4 years. N_2O is removed by ultraviolet (UV) photolysis and attack by $O(^1D)$, which is an electronically excited oxygen atom generated in the stratosphere by O_3 photolysis at wavelengths less than 320 nm. N_2O has a lifetime of about 100 years and is essentially inert in the troposphere.

Lifetimes vary with location and time. In the planetary boundary layer (PBL)—that part of the atmosphere interacting directly with the surface of the earth and extending to about 2 km—lifetimes tend to be short; below a temperature inversion, dry deposition can rapidly remove reactive species such as NH_3. Table 3-4 summarizes typical lifetimes in the PBL for species of interest in this report.

Above the PBL, in the troposphere where wind speeds are higher, temperatures lower, and precipitation is less frequent, the lifetime and range of a pollutant may be much greater. Convection transports short-lived chemicals from the PBL to the free troposphere, where they are diluted by turbulent mixing and diffusion. For key atmospheric species involved in nonlinear processes, such as NO and cloud condensation nuclei (CCN), convection can transform local air pollution problems into regional or global problems.

Spatial Dynamics

Concentrations in the atmosphere depend on the rates and spatial extents of emissions at the surface and on transport, mixing, and reaction in the lower atmosphere. Micrometeorological tools are being developed to support both forward

TABLE 3-4 Typical Lifetimes in the Planetary Boundary Layer for Pollutants Emitted from Animal Feeding Operations

Species	Lifetime
NH_3	~1-10 days
N_2O	100 years
NO	~1 day
CH_4	8.4 years
VOCs	Hours to months, depending on the compound
H_2S	~1 day
PM	1-10 days, depending on particle size and composition
Odor[a]	—

[a]Odor, which results from a mixture of NH_3, VOCs, and H_2S, is an olfactory response to what is often a complex mixture of compounds; it decreases with time after an emission event in response to dispersion (dilution), deposition, and chemical reactions.

calculations, where given emission rates are used to estimate downwind concentration fields, and inverse calculations, where measured concentration fields are used to estimate emission rates. In either of these approaches, the tools must account for the characteristics of the surface and their effects on the flow field, and the effects of regional meteorology. The magnitude of surface roughness, in concert with large-scale pressure gradients, affects the vertical mixing in the lower atmosphere (i.e., dispersion), and the spatial variability of roughness affects the geometry of the mean wind field (i.e., the trajectory of plumes). The strength of surface heating affects the stability of the air column, with convective (unstable) conditions leading to enhanced vertical mixing and temperature inversion (stable) conditions leading to inhibited vertical mixing. From considerations of the effects of PBL stability on the vigor of mixing, it is clear that extreme local pollution episodes generally occur under conditions that combine low horizontal wind speeds—as is often the case when a high-pressure ridge dominates the synoptic-scale weather—with relatively stable temperature profiles (e.g., cooler surface temperatures with warmer air aloft). This results in slow advection horizontally from the source and inhibited mixing vertically, contributing to high surface concentrations (with the emitted mass spread over a relatively small volume of air). A strong temperature inversion (temperature increasing rapidly with elevation) also prevents transport of pollutants to the free troposphere. Consequently, local concentrations are generally highest when ground-level inversions are strongest. A variety of processes, including subsidence, radiation, and advection, can cause inversions. Local orographic conditions, such as lying in a valley, can exacerbate inversions and dictate the wind fields. A detailed discussion is beyond the scope of this report. (A discussion of meteorological effects on concentrations of carbon monoxide, which results mostly from vehicle emissions, can be found in NRC, 2002d.)

Gaussian dispersion models, as commonly used in a regulatory context, have the advantage of providing simple analytical results. However, these models rely on a suite of restrictive conditions that severely limit their robustness under real-field conditions. The transport and mixing from agricultural production facilities are complicated by complex land surface features and transient meteorological conditions. Specifically, the manner in which the patterns of surface roughness, topography, and heating control the trajectory and dispersion of concentration plumes is not captured by these types of models. Observations have demonstrated the failures. Thus, inverse modeling approaches (in which atmospheric concentration measurements are used to estimate the underlying emissions) based on Gaussian models are of limited use in the context of estimating emissions from AFOs.

Progress is being made with a combination of field measurements and Eulerian modeling approaches. Complex landscapes lead to complex solution spaces, thus dictating the need for longer-term observations to characterize the transport-mixing and to identify its controls. Moreover, there is need for increased effort with multidimensional observations, such as with scanning lidar (light detection and ranging) that can characterize evolving plume geometries. (A lidar is a device similar to radar except that it emits pulsed laser light rather than microwaves.) These data sets provide the basis for the construction and testing of multiscale Eulerian modeling frameworks; coarse Eulerian mesoscale models provide regional meteorological forcing and finer scale nested Eulerian models predict plume characteristics over local surface features. However, there has not been a widespread use of these advanced modeling techniques in the AFO regulatory context.

The complexities of the various kinds of air emissions and the temporal and spatial scales of their distribution make direct emission measurements at the individual AFO level generally impractical and cost prohibitive other than in a research setting. Relatively straightforward methods for measuring emission rates by measuring airflow rates and the concentrations of emitted substances are often not available. Flow rates and pollutant concentrations may be available for confined animal housing with forced ventilation, but usually not for emissions from lagoons or soils. An increased research effort on measurement technologies and three-dimensional modeling for flow and transport over complex terrain, with a further focus on stable PBL conditions, is needed to close the gap between the available tools (which presently include restrictive idealized assumptions) and field situations of interest.

POTENTIAL IMPACTS

This section of the report addresses the environmental and human health impacts of materials emitted from AFOs. The impacts occur on a variety of scales, depending on the species (Table ES-1).

Health Effects

Air emissions from AFOs beyond the property line are partly of concern because of their possible effects on human health. The human health effects of the various substances vary with exposure (concentration × time) of humans. The health effects of the substances noted in the following text are based on known levels of exposure; however, there is little scientific evidence that exposures of humans outside the AFOs themselves have significant effects on human health because the concentrations are usually below threshold levels. This may not be the case within the boundary of the AFO and especially in enclosed animal housing. Most of the concern with possible health effects of air emissions from AFOs focuses on ammonia, hydrogen sulfide, and particulate matter. Odor is also discussed in this section because hydrogen sulfide is an important odor-causing substance. Although the evidence of human health effects outside of AFOs is limited, the committee believes additional research may be warranted.

Ammonia

The health effects of ammonia have long been recognized, and the scientific literature on them is extensive. The most recent toxicologic profile for ammonia, published in 1990 by the Agency for Toxic Substances and Disease Registry (ATSDR) of the Centers for Disease Control and Prevention, summarized the human health effects of inhalation (see Tables 3-5 and 3-6). (Ingestion and dermal exposure were also considered, but are not relevant here.) Effects are listed at the lowest concentration at which they were observed.

Ammonia has a strong, sharp, characteristic odor that many people find objectionable. The odor is generally detectable at concentrations greater than 50 parts per million (ppm), so harmful exposures are likely to be detected early enough for the exposed person to take evasive action. Tables 3-5 and 3-6 show that there is little likelihood of even minimal health effects of long-term exposure to ammonia at concentrations less than 0.3 ppm, and that even concentrations up

TABLE 3-5 Short-Term Exposure[a] to Ammonia

Concentration (ppm)	Length of Exposure	Description of Effects
0.5		Minimal risk level
50	Less than 1 day	Slight, temporary eye and throat irritation and urge to cough
500	30 minutes	Increased air intake into lungs; sore nose and throat
5000	Less than 30 minutes	Kills quickly

[a]Less than or equal to 14 days.

TABLE 3-6 Long-term Exposure[a] to Ammonia

Concentration (ppm)	Length of Exposure	Description of Effects
0.3		Minimal risk level
100	6 weeks	Irritation of eyes, nose, and throat

[a]More than 14 days.

to 50 ppm for a day do not have serious health consequences. Ammonia is also a precursor to secondary ammonium nitrate aerosol, whose health effects are discussed below.

Hydrogen Sulfide

Hydrogen sulfide is a colorless gas with a strong and generally objectionable rotten egg odor, detectable at concentrations down to 0.5 parts per billion (ppb). Paradoxically, most people cannot smell H_2S at concentrations greater than 100 ppm. Most of the data on H_2S toxicity are in the form of no observed adverse effects levels (NOAEL) or lowest observed adverse effect levels (LOAEL). These are, respectively, the highest concentrations at which no adverse effects were noted in the populations examined and the lowest concentrations at which such effects were in fact observed. A threshold of sorts may be presumed to exist at some point between these two exposure concentrations for a specific outcome in the population studied. Whether that threshold is well determined and whether it applies to other populations are generally matters of scientific judgment.

The toxicologic profile of H_2S is described below. ATSDR (1999) summarized the relationships between exposure and health outcomes as follows; data are for humans unless otherwise noted:

Acute Effects

Death. Human studies were not reported, but mortality was 100 percent in small groups of rats exposed for 3 minutes to 1655 ppm, 12 minutes to 800 ppm (males), or 4 hours to 500-700 ppm (males). The LD-50 (the dose for 50 percent mortality) was estimated to be 587 ppm for rats exposed for two hours or 335 ppm for six hours.

Respiratory, Cardiovascular, and Metabolic Effects. The NOAEL for humans ranged from 2 to 10 ppm for one exposure of 15 minutes to two exposures of 30 minutes each.

Immunological Effects. The NOAEL was 50 ppm for rats exposed for four hours.

Neurological Effects. Headaches were reported in 3 of 10 asthmatics exposed to 2 ppm H_2S for 30 minutes. Biochemical changes were found in the brains of guinea pigs exposed to 20 ppm for one hour per day for 11 days, although the NOAEL for impaired performance in rats was 100 ppm for two hours.

Chronic Effects

No human studies were reported, and no animal studies reported on deaths from longer exposure. The NOAEL for exposures up to 90 days seemed to be 80 ppm for most systemic end points in rats and mice, and 8.5 ppm in pigs.

Particulate Matter

Particles are highly complex in terms of size, physical properties, and composition, and there may be important synergistic effects with gases in the air. Particles—PM10 and especially PM2.5—have been linked to health effects found on at least two time scales (Schwartz, 1994). Populations with long-term exposure to heavier loads of particles have higher rates of both total mortality and mortality from major cardiovascular diseases, as well as increased rates of morbidity, expressed primarily as hospital admissions. The magnitude of the effect varies with location and may not be present in every population. A second type of effect, seen in time series studies within a single population, is short-term correlation of PM concentrations with mortality and morbidity, generally within 0-3 days of exposure to above-average concentrations.

Most observers are convinced that these correlations reflect cause and effect rather than the effects of co-pollutants and other confounders, but many also believe that it is not just the particulate nature of the particles that does the damage. Damage seems to be most intense with the smallest particles (less than 2.5 μm or PM2.5), which can be composed of elemental carbon, adsorbed complex organic molecules, heavy metals, bioaerosols, acid aerosols, ammonium nitrate, and other materials. Substantial research interest is now directed at elucidating the aspects of PM that cause various health effects. It is by no means clear that the hazards of rural PM can be inferred from research focused on urban PM.

Odor and VOCs

Many of the complaints about AFOs are generated by odor. As classes of compounds, odor and VOCs can be considered together. In terms of their health and environmental effects, some of the VOCs may irritate the skin, eyes, nose, and throat. They can also be precursors to the formation of tropospheric ozone and PM2.5. Odorous VOCs can stimulate sensory nerves to cause neurochemical changes that might influence health by compromising the immune system. However, the effects of air emissions from AFOs on public health are not fully understood or well studied. Greater mood disturbance (Schiffman et al., 1995) and increased rates of headaches, runny nose, sore throat, excessive coughing, diarrhea, and burning eyes have been reported by persons living near swine operations in North Carolina (Wing and Wolf, 2000). Thu et al. (1997) observed similarities between the pattern of symptoms among community residents living near large swine operations and those experienced by workers. Caution must be exercised in interpreting the studies because environmental exposure data were not reported.

Environmental Impacts

The environmental impacts of the nitrogen-, carbon-, and sulfur-containing species emitted from AFOs are well known and thus are reviewed only briefly here. Nitrous oxide and methane are radiatively active and contribute to the greenhouse effect in the troposphere (IPCC, 2001; NRC, 2001b). Some VOCs participate in atmospheric photochemical reactions, while others play an important role as heat-trapping gases (King, 1995). A large portion of N_2O is transferred to the stratosphere where it has the potential to contribute to stratospheric ozone depletion. Nitric oxide and ammonia also have environmental impacts. Nitric oxide contributes to increased concentrations of O_3 in the troposphere and can result in decreased productivity of crops and terrestrial ecosystems (Ollinger et al., 2002). Ammonia in the atmosphere can react with sulfuric and nitric acids to form ammonium sulfate and nitrate aerosols. It plays an important role in the direct and indirect effects of aerosols on radiative forcing and thus global climate change (Seinfeld and Pandis, 1998; Penner et al., 2001).

In addition to these effects of emitted NO and NH_3, both species also contribute to a wide variety of other environmental impacts as they are converted to other chemical species and cycle through environmental reservoirs. Referred to as the "nitrogen cascade," the sequential transfer of reactive nitrogen through environmental systems results in environmental changes as nitrogen moves through each system or is temporarily stored (Galloway and Cowling, 2002). The cascade is illustrated in Figure 3-2.

Nitrogen is converted by human activity from molecular nitrogen (N_2) to reactive NO_x and NH_x (NH_x is used here to mean the total of NH_3 and NH_4^+). New NO_x is produced primarily by combustion processes (energy production), while NH_3 is produced by the Haber-Bosch process to make fertilizer used to increase food production. About half of the nitrogen in fertilizer applied to global agroecosystems is incorporated into crops and used for human and livestock consumption (Smil, 1999, 2001); the other half is transferred to the atmosphere as NH_3, NO, N_2O, or N_2 or lost to aquatic ecosystems, primarily as nitrate. Ammonia from agroecosystems can follow a variety of pathways, resulting in a sequential series of impacts; the same is true for NO_x. In sequence, an atom of nitrogen (in NO_x) can first increase tropospheric ozone, then produce small particles that decrease atmospheric visibility, and then increase acidity in precipitation. Following deposition to terrestrial ecosystems, that same nitrogen atom can increase soil acidity, decrease biodiversity, and increase or decrease ecosystem productiv-

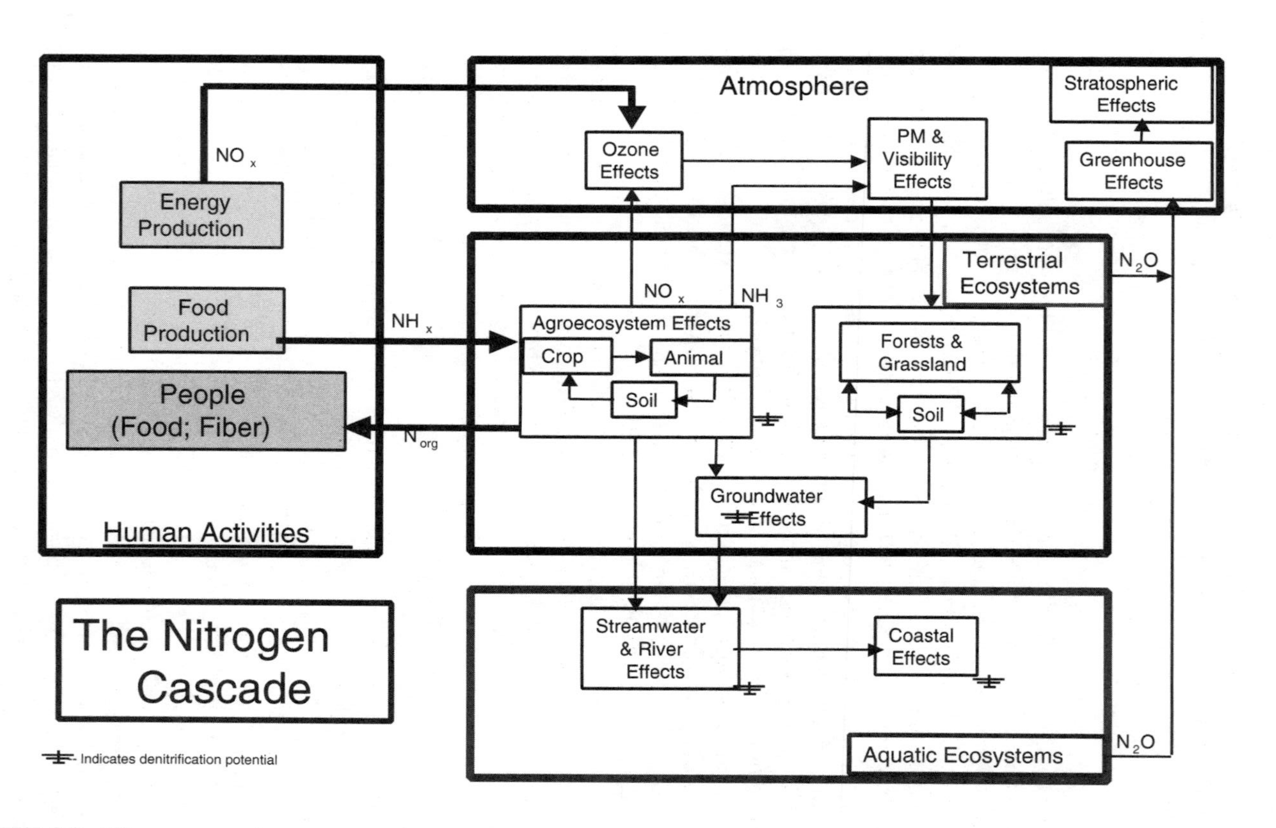

FIGURE 3-2 Nitrogen cascade.
SOURCE: Galloway and Cowling (2002).

ity. If discharged to aquatic ecosystems, it can increase surface water acidity and lead to coastal eutrophication. If the nitrogen atom is converted to N_2O and emitted back to the atmosphere, it can first increase greenhouse warming and then deplete stratospheric ozone. As Figure 3-2 illustrates, an atom of reactive nitrogen (Nr), can enter the cascade at different places. An important characteristic of the cascade is that once an Nr atom enters, its source (e.g., fossil fuel combustion, fertilizer production) becomes irrelevant—except for different types of control strategies that may be employed.

The emissions of H_2S from AFOs generally make a minor contribution to the sulfur burden of the atmosphere relative to the SO_2 from the burning of fossil fuels. Where there are few other sulfur sources however, H_2S can contribute significantly to the PM2.5 burden.

Particulate matter in the atmosphere decreases visibility. The primary environmental and ecologic effects of particles are related to haze, which is caused by the suspended aerosols that both absorb and scatter light. The primary constituents of concern are carbonaceous materials (absorption) and aerosols containing sulfates and nitrates (scattering). Even though the atmosphere naturally limits our ability to distinguish an object from background by Rayleigh scattering (when the size of the aerosol is much less than the wavelength of visible light), Mie scattering (when the size of the aerosol is approximately equal to the wavelength of visible light) can greatly decrease visibility.

FINDING 2. Air emissions from animal feeding operations are of varying concern at different spatial scales, as shown in Table 3-7.

RECOMMENDATION: These differing effects, concentrations, and spatial distributions lead to a logical plan of action for establishing research priorities to provide detailed scientific information on the contributions of AFO emissions to potential effects and the subsequent implementation of control measures. USDA and EPA should first focus their efforts on the measurement and control of those emissions of major concern.

FINDING 3. Measurement protocols, control strategies, and management techniques must be emission and scale specific.

RECOMMENDATIONS:

- **For air emissions important on a global or national scale (i.e., ammonia and the greenhouse gases methane and N_2O), the aim is to control emissions per unit of production (kilograms of food produced) rather than emissions per farm. Where the environmental and health benefits outweigh the costs of mitigation it is important to decrease the**

TABLE 3-7 Committee's Scientific Evaluation of the Potential Importance of AFO Emissions at Different Spatial Scales

Emissions	Global, National and Regional	Local—Property Line or Nearest Dwelling	Primary Effects of Concern
NH_3	Major[a]	Minor	Atmospheric deposition, haze
N_2O	Significant	Insignificant	Global climate change
NO_x	Significant	Minor	Haze, atmospheric deposition, smog
CH_4	Significant	Insignificant	Global climate change
VOCs[b]	Insignificant	Minor	Quality of human life
H_2S	Insignificant	Significant	Quality of human life
PM10[c]	Insignificant	Significant	Haze
PM2.5[c]	Insignificant	Significant	Health, haze
Odor	Insignificant	Major	Quality of human life

[a]Relative importance of emissions from AFOs at spatial scales based on committees' informed judgment on known or potential impacts from AFOs. Rank order from high to low importance is major, significant, minor, and insignificant. While AFOs may not play an important role for some of these, emissions from other sources alone or in aggregate may have different rankings. For example VOCs and NO_x play important roles in the formation of tropospheric ozone; however the role of AFOs is likely to be insignificant compared to other sources.

[b]Volatile organic compounds.

[c]Particulate matter. PM10 and PM2.5 include particles with aerodynamic equivalent diameters up to 10 and 2.5 μm, respectively.

aggregate emissions. In some geographic regions, aggregate emission goals may limit the number of animals produced in those regions.

- **For air emissions important on a local scale (H_2S, particulate matter, and odor), the aim is to control ambient concentrations at the farm boundary and/or nearest occupied dwelling. Standards applicable to the farm boundary and/or nearest occupied dwelling must be developed.**
- **Monitoring should be conducted to measure concentrations of air pollutants with possible health concern at times when they are likely to be highest and in places where the densities of animals and humans, and typical meteorological conditions, are likely to result in the highest degree of human exposure.**

FINDING 4. There is a general paucity of credible scientific information on the effects of mitigation technology on concentrations, rates, and fates of air emissions from AFOs. However, the implementation of technically and economically feasible management practices (e.g., manure incorporation into soil) designed to decrease emissions should not be delayed.

RECOMMENDATION: Best management practices (BMPs) aimed at mitigating AFO air emissions should continue to be improved and applied as new information is developed on the character, amount, and dispersion of these air emissions, and on their health and environmental effects. A systems analysis should include impacts of a BMP on other parts of the entire system.

SUMMARY

This chapter described the various constituents of air emissions that are of primary interest and the factors that determine emissions rates and dispersion as they affect atmospheric concentrations. The effects of temporal and spatial factors are described along with the complexities of modeling the flow of emissions over complex terrain as they affect field measurements. Potential impacts of these air emissions on human health and the environment are described and put in the context of expected rates and concentrations of emissions from animal feeding operations.

4

Measuring Emissions

INTRODUCTION

Emissions from animal feeding operations (AFOs) have local, regional, and global impacts. The committee was charged by the U.S. Environmental Protection Agency (EPA) and the U.S. Department of Agriculture (USDA) to provide recommendations on the most promising science-based methodologies and modeling approaches for estimating and measuring these emissions. This implies a desire by EPA and USDA to quantify and perhaps limit these emissions and to evaluate decreases made possible by abatement strategies and management practices.

Air quality in the United States is maintained through the adoption of both National Ambient Air Quality Standards (NAAQS) and air pollutant source emission standards that are applied to numerous sources. The latter implies that the emission rates of specific air pollutants from each source type are known. If an airshed does not meet the NAAQS, knowledge of the emission rates from the various sources in the airshed allows the development of strategies to improve air quality by targeting important sources for control. While this approach to regulation may be straightforward for some industrial operations, livestock feeding operations pose especially complicated issues for estimating and measuring air emissions.

AFOs are complex physical, chemical, and biological systems. Feeding, digestion, excretion, and animal and site activity show diurnal, seasonal, and life cycle variations. Once feces and urine leave the animal, they undergo a variety of processes that are driven by wind, temperature, moisture, and microbial metabolism. Factors such as pH and the availability of oxygen affect the communities of microorganisms that are going through their own life cycles and are responsible for transforming the nonvolatile compounds in feed, water, and manure into volatile compounds such as ammonia (NH_3), nitrous oxide (N_2O), nitric oxide (NO), methane (CH_4), and volatile organic compounds (VOCs), as well as hydrogen sulfide (H_2S) and other odor-causing compounds.

A complication in measuring emissions from AFOs is that most emissions are released from area sources such as cattle feedlots, wastewater lagoons, or the land to which manure or lagoon liquid is applied, rather than from a few discrete point sources (e.g., animal house exhaust fans). Spatial and temporal variation exists. Measuring area emission rates often depends on measuring atmospheric concentrations and characterizing the micrometeorology or using atmospheric dispersion models to back-calculate the emission rates that gave the concentrations observed. Measuring emission rates from animal housing with forced ventilation is relatively easier; one measures concentrations and ventilation airflow rates. The variability in atmospheric concentrations possible near an area source is illustrated by measurements of ammonia shown in Figure 4-1. Over a period of about half an hour, the average NH_3 concentration near a dairy wastewater lagoon varied from about 10 to 700 ppb (parts per billion). This variability (a factor of 70) was due primarily to variable wind speed and direction during the measurement period.

Although direct measurement of off-property impacts of the various emissions from every AFO is not practical, there is a need for an approach that can be used by local, state, or federal agencies to estimate emissions from individual AFOs. The overall air quality management goal is to limit emissions to concentrations that will not lead to exceedances of the NAAQS for criteria pollutants or other regulatory limits described in Chapter 6.

AFO emissions have impacts on several spatial and temporal scales. Greenhouse gases such as nitrous oxide and methane, which have long atmospheric half-lives and are transported for long distances, have global rather than local or regional effects; their annual emissions from U.S. agriculture are important but their local or regional concentrations are not. The fraction of these gases from AFOs is of some concern because some kinds of controls may be applied more efficiently to large sources than to smaller ones.

Primary particulate matter and odors are of concern mostly to individuals near the emission sources. What are important for them are not annual totals, but ambient concentrations averaged over short periods of time (typically 1 to 24 hours). These concentrations depend not only on short-term emission rates, but also on meteorological conditions at the time, including wind speed and direction, atmospheric stability, and precipitation. Some pollutants act at a variety of scales. Ammonia and H_2S contribute to short-range odor and toxicity, but react in the atmosphere to form secondary fine particulate matter (PM2.5) dispersed over a regional scale; VOCs contribute to odor and also react with nitrogen oxides (NO_x) in the presence of heat and sunlight to form tropospheric ozone (O_3), another regional problem. For air pollutants that can have adverse human health

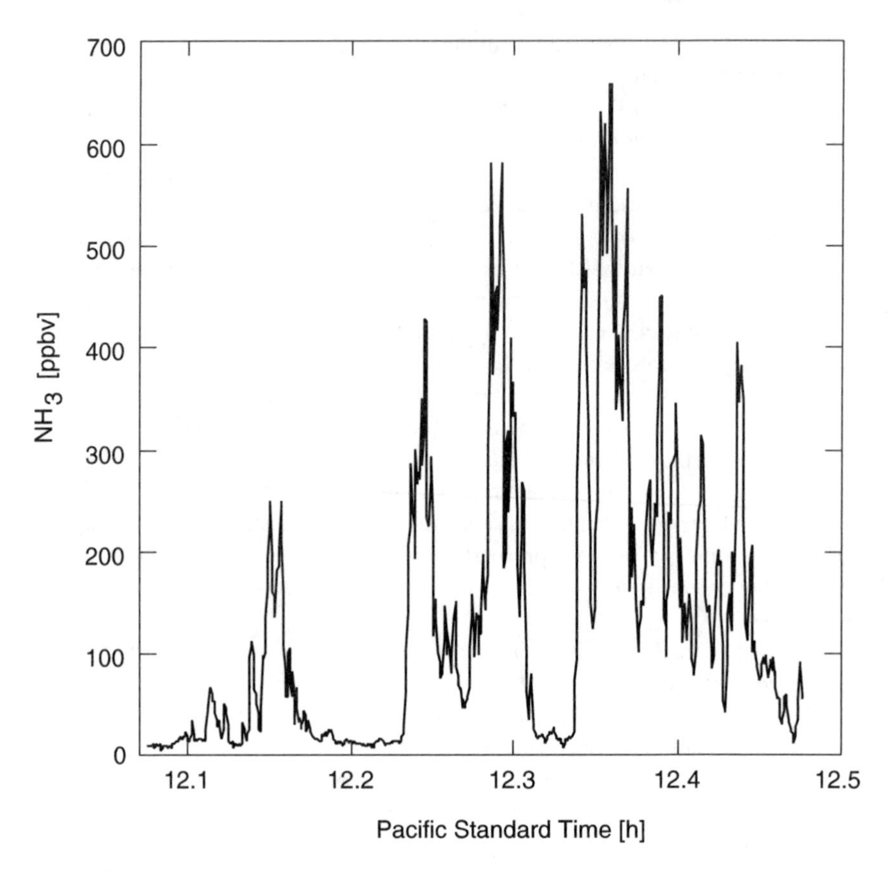

FIGURE 4-1 Ammonia concentrations (averaged over a 68-m path) measured near a dairy wastewater lagoon. Method used was an ultraviolet differential optical absorption spectroscopy technique with a detectability limit of 1 ppb and a time resolution of 0.6 second.
SOURCE: Reprinted from Mount et al., 2002, with permission from Elsevier Science.

effects, the time course of concentrations inhaled is important; for PM, health effects also depend on the physical, chemical, and biological characteristics of the particulate material. Concentrations of odorants are important, but measurement and understanding of their health effects are still in their infancy.

CALIBRATION, SAMPLING, AND CONCENTRATION MEASUREMENTS

The substances of interest in this report run the gamut from the most accurately measured atmospheric trace gases, with a long history of precise calibration standards, to the most challenging biologically active substances, for which accurate measurement remains a research topic. Measurement methods and calibration standards for methane, nitric oxide, and nitrous oxide are among the most thoroughly studied. Observations can be made with well-characterized methods and with low uncertainty relative to natural variability. Controlled intercomparisons of various analytical techniques under field conditions have led to uniform international calibration standards and the identification of reliable instruments. In ambient air under a variety of weather conditions and for a broad range of concentrations, CH_4 and N_2O can be measured within 1 percent and NO within 15 percent (Fehsenfeld et al., 1987). In the case of odor, no formal effort to standardize methodology has occurred in the United States.

Calibration Gases

The more stable gases, CH_4, N_2O, NO, and some nonmethane hydrocarbons (NMHCs) can be stored as compressed gas mixtures. The mixing ratios of these gases remain stable for years if the cylinders are stored and handled properly, so these mixtures can provide reliable calibration standards with absolute uncertainty well below 10 percent.

NIST Standard Reference Materials

Calibration gas standards of CH_4, NO, and N_2O in air or in nitrogen (certified at concentrations of approximately 5 to 40 ppm [parts per million]) are obtainable from the Standard Reference Material (SRM) Program of the National Institute of Standards and Technology (NIST), formerly the National Bureau of Standards (NBS), in Gaithersburg, Maryland. These SRMs are supplied as compressed gas mixtures at about 135 bar (1900 pounds per square inch [psi]) in high-pressure aluminum cylinders containing 800 L (liters) of gas at standard temperature and pressure, dry (STPD) (NBS, 1975; Guenther et al., 1996). Each cylinder is supplied with a certificate stating concentration and uncertainty. The concentrations are certified to be accurate to ±1 percent relative to the stated values. Because of the resources required for their certification, SRMs are not intended for use as daily working standards, but rather are to be used as primary standards against which transfer standards can be calibrated.

NIST Traceable Reference Materials

Calibration gas standards of CH_4, N_2O, and NO in air or nitrogen (N_2), in the concentrations indicated above, are obtainable from specialty gas companies. Information as to whether a company supplies such mixtures is obtainable from the company or from NIST's SRM Program. These NIST Traceable Reference Materials (NTRMs) are purchased directly from industry and are supplied as compressed gas mixtures at approximately 135 bar (1900 psi) in high-pressure aluminum cylinders containing 4000 L of gas at STPD. Each cylinder is supplied with a certificate stating concentration and uncertainty. The concentrations are certified to be accurate to within ±1 percent of the stated values (Guenther et al., 1996).

More recent work (Phillips et al., 2001) recommends the annular denuder sampling with on-line analysis rotary annular denuder technique for continuous measuring of ammonia concentrations from AFOs. Wet chemistry methods (acid bubblers or passive "ferm tube" samples) are generally acceptable but cannot measure continuously. For determining emission rates that require measurement of ventilation rate, Phillips et al. (2001) suggests the use of fan-wheel anemometers or a tracer gas method using sulfur hexafluoride for force- and naturally-ventilated buildings, respectively.

MEASURING CONCENTRATIONS AND COMPOSITIONS

Ammonia

For ambient (outside) air, several reviews and intercomparisons of ammonia detection techniques have been published (Harrison and Kitto, 1990; Wiebe et al., 1990; Williams et al., 1992a). Results of the intercomparisons have been inconclusive. Interference from particles and problems with sampling and conversion efficiency (including temperature and humidity dependence) complicate measurements. The long-term stability of ammonia in compressed gas mixtures has not been demonstrated, but calibration standards can be produced using permeation devices.

Samples collected using both denuders and filters can be analyzed by aqueous-phase colorimetry or ion chromatography to measure NH_3 and ammonium ion (NH_4^+). Diffusion denuders coated to remove specific gases rely on the relatively rapid diffusion of gases compared to particles to remove the gases while allowing the particles to pass through the denuder. With sequential gas samplers, for example, air is drawn through two parallel channels, each containing a quartz filter followed by a cellulose filter impregnated with citric acid. In front of one sampler is an NH_3 denuder made of citric acid-coated parallel tubes, which remove vapor-phase ammonia but not particulate-phase ammonium. The impregnated filters collect ammonium and ammonia (as ammonium ion) quantitatively. Ammonium is extracted from the filters into water and measured via colorimetry or ion chromatography. Gaseous ammonia is calculated from the difference between the two channels. Chow et al. (1996) report that the analytical uncertainty in a single measurement is typically ±10 percent for a measured value that is more than 10 times below detection limits, but this accuracy has not been confirmed by independent analysis. The response time depends on the concentration, varying from days to hours.

Ammonia can be converted to NO on hot stainless steel in the presence of oxygen and the NO detected via chemiluminescence, as described below. Conversion can take place either continuously or following collection onto a denuder. This technique suffers from potential interference from other substances, such as nitrogen dioxide (NO_2) and nitric acid (HNO_3), that are also converted to NO. To avoid such interferences, the system can be zeroed with an ammonia-specific scrubber such as an acid-coated denuder tube. Response times vary from minutes to hours; absolute accuracy and precision are unknown.

Ammonia can be dissociated photolytically and the NH fragment detected with photolytic fragmentation laser-induced fluorescence (PF-LIF). Heating by a tuned carbon dioxide (CO_2) laser followed by interferometry to measure ammonia has also been reported (Owens et al., 1999). These techniques are labor intensive and expensive, but sensitive and fast relative to filter techniques.

In an intercomparison conducted in a cold dry environment in Colorado (Williams et al., 1992b), the photofragmentation and denuder systems agreed well, within 15 percent in some cases. Problems with sampling and gas-to-particle conversion caused some filter techniques to differ from other methods by about 35 percent. All techniques generally agreed to within a factor of two for fog-free conditions. Fourier transform infrared spectrometry has been used with some success to measure trace gas concentrations in swine facilities (Childers et al., 2001), although the conversion to emission fluxes has met with only very limited success. Tunable diode laser techniques can also be applied with sensitivity levels of several parts per billion. Older techniques involve use of ferm tubes, absorption flasks, filter badges, and shuttles.

Major problems associated with NH_3 detection and analyses are related to emissions of NH_3 to instrument inlet surfaces. Many methods also rely on integrated sample collection. Mount et al. (2002) recently described a new technique for measuring atmospheric NH_3 in real time, using differential optical absorption spectroscopy (DOAS). Their method avoids a collecting medium and thus circumvents the common problem of losses to inlet surfaces. They reported a detection limit on the order of 1 ppb. This method has not yet been tested in intercomparison studies.

Nitrous Oxide

For nitrous oxide, gas chromatography (GC) followed by electron capture detection (GC-ECD), tunable diode laser spectroscopy (TDLS), and Fourier transform infrared spectroscopy (FTIR) have demonstrated accuracy well above the natural variability of the emitting systems. Esler et al. (2000) report N_2O measurement precision of 1 percent, which is inadequate for long-term global trend analysis, but more than adequate for emission measurements. GC-ECD has been used successfully for some time (e.g., Robertson et al., 2000). Intercomparisons have indicated agreement of standards to within 1 percent. These techniques work on both ambient air samples and grab samples stored in canisters.

Nitric Oxide

Nitric oxide is generally measured in situ with instruments based on chemiluminescence. NO at ambient concentrations cannot be reliably stored in grab samples for later analysis; it is too reactive at typical concentrations in ambient air. Chemiluminescence detection of NO involves reaction with excess ozone to produce electronically excited NO_2, which then relaxes to emit a photon. This technique has been investigated extensively in both laboratory and field studies (Fehsenfeld et al., 1987; Crosley, 1996). It compares well with laser-induced fluorescence and is generally accepted to be reliable (at concentrations relevant to emissions studies) to better than 15 percent with 95 percent confidence. Response times are on the order of 1 minute.

Methane

Methane is measured most commonly with GC followed by flame ionization detection (FID). Interagency intercomparisons indicate agreement to well within 1 percent (Masarie et al., 2001). The technique works on both ambient samples and air stored in canisters. TDLS has also proven successful for monitoring CH_4 (Billesbach et al., 1998).

Volatile Organic Compounds

Hydrocarbons can be measured with gas chromatography followed by flame ionization detection (GC-FID). Detection by mass spectroscopy is sometimes used to confirm species identified by retention time (Westberg and Zimmerman, 1993; Dewulf and Van Langenhove, 1997). Preconcentration is typically required for less abundant species. In an intercomparison conducted with 16 components among 28 laboratories, agreement was typically tens of percent (Apel et al., 1994). In a more recent intercomparison (Apel et al., 1999), 36 investigators from around the world were asked to identify and quantify C_2 to C_{10} hydrocarbons (HCs) in a mixture in synthetic air. Calibration was based on gas standards of individual compounds, such as propane in air, and a 16-compound mixture of C_2 to C_{16} *n*-alkanes, all prepared by NIST and certified to ±3 percent. The top-performing laboratories, including several in the United States, obtained agreement of generally better than 20 percent for the 60 compounds. Intercomparison of NMHCs in ambient air has yet to be reported. Measurement of other volatile organic compounds such as those containing nitrogen, oxygen, or sulfur remains a focus of ongoing research; a summary of these techniques is beyond the scope of this report.

Hydrogen Sulfide

Wet chemistry techniques, gas analyzers, and monitors are available for measuring concentrations of hydrogen sulfide in air. The wet chemistry technique involves the collection of sulfides in heavy-metal salt solutions and later recovery and measurement of sulfide using wet chemistry methods. The heavy-metal salt solutions, when exposed to H_2S, form insoluble metal sulfide precipitates (Barth and Polkowski, 1974; Elliott et al., 1978). Zinc acetate, mercuric chloride ($HgCl_2$), and mercuric cyanide ($Hg[CN]_2$) solutions are common. The concentration of H_2S in air was calculated by the mass of it collected in the solution divided by the volume of air, which is determined from the flow rate of air that travels through the collection impingers and the time period of collection. Wet chemistry techniques tend to be time consuming and may experience loss of sulfide due to incomplete recovery of sulfide from the precipitates.

When the concentration of H_2S in the air sample is below the detection limit of GC, concentrating procedures, such as cold traps (Beard and Guenzi, 1983) and adsorbent materials (Elliott et al., 1978), must be used. Gas chromatography provides good dependability and accuracy for gas analysis and has been used by researchers for H_2S measurements (Banwart and Bremner, 1975; Powers et al., 2000). However, the relatively poor portability of GC limits its use mostly to the environments of the research laboratory. Therefore, GC is usually not suited for on-site monitoring unless a mobile laboratory is provided at the measurement site. Portable and durable analyzers and monitors are more desirable for field measurement and monitoring. Three gas analyzers that have been used recently for H_2S concentration monitoring around livestock farms are described below.

For the Zellweger MDA Single Point Air Monitor (SPM) the detection limit is 1 ppb for hydrogen sulfide (2.6 ppm for ammonia). The accuracy is 20 percent. The SPM uses a paper tape treated with a dry reagent medium to collect and analyze the gas. Upon exposure to the target gas, the paper tape changes color in direct proportion to the gas concentration in the air sample. A photo-optic system within the SPM measures the color intensity range and determines the gas concentration based on 15-minute averages. Bicudo et al. (2002) used this instrument for measuring ambient H_2S concentrations near swine barns and manure storages.

The Portable Jerome Hydrogen Sulfide Analyzer (Model 631-X, Arizona Instrument, Phoenix, AZ) has an accuracy of 6-8 percent depending on the concentration of H_2S to be measured. This instrument uses a gold film sensor for detection and measurement. It has been used recently by several researchers for monitoring ambient H_2S concentrations around livestock farms (Wood et al., 2001). It was noted that the sensor was affected by other sulfide compounds. However, due to the lower response factors of the sensor to them, the Jerome meter is considered to be capable of providing quantitative detection of H_2S.

The TEI Model 45C H_2S Analyzer has an H_2S-to-SO_2 (sulfur dioxide) converter coupled to a pulsed fluorescence SO_2 analyzer. The detection limit is 0.5

ppb. Ni et al. (2002a) and Zahn et al. (2002) used this instrument for H_2S analysis in swine buildings and over a swine wastewater lagoon.

Based on researchers' reports and manufacturers' information, the Portable Jerome Hydrogen Sulfide Analyzer and the TEI Model 45C H_2S Analyzer are considered good choices of instrumentation for measuring and monitoring ambient H_2S concentrations at livestock farms.

The experience of researchers with field measurement and monitoring of H_2S is scarce. The consistency and accuracy of H_2S analyzers have to be better understood, and proper usage and calibration procedures must be developed.

Particulate Matter

Measurement technologies for particulate matter (PM) are profoundly affected by the complexity of PM emissions. PM is not a single well-defined entity such as N_2O or CH_4. The actual constituents vary, as do particle sizes, depending on geographical location and meteorological conditions. Fresh particles from urban sources can be quite active; the aerosol chemistry in polluted, urban airsheds is intricate and dynamic. Combustion sources such as motor vehicles and boilers release volatile and semivolatile organic compounds, some of which quickly condense to form very small particles. These ultrafine particles serve as condensation nuclei upon which other organic and inorganic vapors condense, thus growing the particles in the proximity of their sources. As it is transported downwind, the aerosol mix "ages" as numerous other processes take place. Organic vapors condense on organic and other types of aerosol particles, including soil and inorganic particles. Gaseous ammonia reacts with gaseous nitric acid to form particulate ammonium nitrate. Particles containing sulfuric acid form from the oxidation of gaseous SO_2; these react with other gases, including ammonia. Smaller particles tend to agglomerate into larger ones; photochemistry can take place in some. The primary particles emitted directly from the combustion sources and the secondary particles generated from these atmospheric reactions and from condensation growth tend to be smaller than 1 μm in aerodynamic diameter ("fine particulate matter"). Thus, the nature of PM depends not only on the source, but also on other co-pollutants, chemical reactions, and distance from the source.

By comparison, PM from animal feeding operations tends to contain a significant fraction of mechanically generated material such as soil, dried manure, and so forth—material that is typically larger than 1 μm in aerodynamic diameter ("coarse particulate matter"). In many rural areas with AFOs, especially in arid and semiarid regions, there is a relatively large mass concentration of coarse particulate matter, compared to fine. Air in rural agricultural areas may also carry a significant load of fine PM from the reaction of gaseous ammonia with other chemical compounds such as nitric acid. In addition, it has been speculated that the particulate organic material that exists in rural areas may contain bioaerosols such as toxins, allergens, viruses, bacteria, and fungi. To date, studies of bioaerosols in swine barns have been limited, in part because there are no currently accepted standard methods for bioaerosol measurement, and in part because no sampler has been fully characterized for bioaerosol collection efficiency.

Chemical changes may take place in the rural atmosphere; however, the chemistry is likely to be less complex than that in the urban atmosphere, especially with respect to particulate organic matter. In any event, the chemical composition of particulate matter in rural areas will be substantially different from that in urban areas. Whatever the complexities, it is by no means clear that the hazards of rural PM can be inferred from research focused on urban PM, but at present both are regulated in the same way.

Atmospheric particulate matter is measured by a variety of techniques, but for regulatory purposes, the primary measures are PM10 and PM2.5. Both PM10 and PM2.5 are "operationally defined" by the measurement technique. PM10 is the particulate matter captured in a size–selective inlet that removes particles with an aerodynamic equivalent diameter (AED) of 10 μm with an efficiency of 50 percent. PM2.5 is similarly defined but for particles with AED of 2.5 μm. The AED is not a true physical dimension of the particle, but rather an equivalent diameter based upon a spherical particle with a density of 1 g/cm^3 with the same settling velocity as the particle in question. In measurements of both PM10 and PM2.5 there is some collection of particles larger than the indicated size and loss of some smaller, since a perfectly sharp cutoff based on particle size is not currently possible.

Both PM10 and PM2.5 samplers collect particles with a variety of sizes, shapes, and compositions. Size-selective sampler inlets are usually based on some type of inertial separation to remove larger, undesired particles. The remaining material is then collected onto some sort of sampling media, typically a Teflon, quartz, or nylon filter substrate. The mass concentration is determined by weighing the filter before and after sampling and dividing the weight increase by the total air volume that passed through the sampler.

Chemical and biological analyses are normally carried out on integrated samples collected over a period of time, typically 24 hours. Other methods are available for measuring PM in real (or semireal) time, but there is currently no continuous single method that measures true particulate mass for all compositions. Complications of current samplers include weight loss from semivolatile materials and weight gain by adsorption of gas-phase substances—including water—during sampling and handling. Continuous methods that rely on optical techniques must be calibrated for each type of particulate matter, since optical properties are a function of particle size and composition. Moreover, particles larger than about 2 μm are not very active optically, making instruments (e.g., nephelometers) based on these methods less reliable for the coarse PM that is often encountered in rural areas with AFOs.

Water contributes significantly to the mass of some PM samples. In integrated sampling, the commonly accepted procedure involves equilibrating filters before and after sampling at a constant specified temperature and relative humidity to control the particle-bound water. In many continuous samplers the problem of particle-bound water, especially at high relative humidity, is dealt with by heating the inlet. This can cause a problem when monitoring for PM2.5 containing a significant fraction of ammonium nitrate, whose heating can cause decomposition back to NH_3 and HNO_3. A similar weight loss problem occurs when heating semivolatile organic compounds causes desorption from the filter.

Problems associated with determining true atmospheric PM concentrations are compounded by the lack of available calibration standards. While there is a NIST standard for urban PM, there is, at this time, no NIST-traceable standard aerosol mix that can be used to calibrate particulate samplers and monitors. It is possible to calibrate the individual components of a PM sampler with NIST or NIST traceable standards, but it is not possible to calibrate the entire PM sampler system. Quality control procedures test the individual components of the measurement and estimate the overall uncertainty using propagation of error analysis. Individual components that can be assigned measures of quality include flow rate, particulate mass, and temperature and pressure. Other quality control procedures include the use of lab and field blanks and duplicate sampling and analyses. There is considerable ongoing effort in the research community to improve and refine the measurement of atmospheric particles and their constituents, on a real-time, size-segregated, and chemically speciated basis.

Odor

Odor is a sensation produced by stimulation of chemoreceptors in the olfactory epithelium in the nose. The substances that produce the sensation, which include NH_3, VOCs, and hydrogen sulfide (H_2S), are called odorants. Zahn et al. (1997) established a significant correlation between air concentrations of VOCs and odor offensiveness in swine production facilities. Odor threshold (OT) refers to the minimum concentration of odorant necessary for perception in a specified percentage of the population, usually 50 percent; it is a statistical value, representing the best estimate from a group of individual scores. If only a single compound produces the odor, odor intensity can be expressed in terms of the concentration (e.g., micrograms per cubic meter) of that compound. However, with odors from animal sources, which are generally caused by complex mixtures of compounds, odor concentration is expressed in terms of odor units (OUs) per cubic meter. A concentration of 100 OU/m^3 means that a given volume of odorous air must be diluted with 100 volumes of odor-free air before it reaches the detection threshold.

In the recently developed European Standard, the unit of measurement for odor concentration is the European Odor Unit per cubic meter (OU_E/m^3; European Committee for Standardization, 2002). An OU_E is defined as the amount of odorant(s) in 1 m^3 of neutral gas at standard conditions that elicits a physiological response from a panel at detection threshold equivalent to that elicited by 123 μg of *n*-butanol (1 European Reference Odor Mass). Researchers in the United States have not formally adopted this as the standard unit of measurement.

Odor Sampling

Proper collection and storage of odor samples prior to presentation to panelists or instruments is important for ensuring the accuracy of the measurements. Samples may be collected for immediate or delayed analysis. In sampling for immediate analysis (also called dynamic sampling), the odor sample is ducted directly to an analyzer, such as an olfactometer, without intermediate storage. In sampling for delayed analysis, a sample is collected and transferred to a sample container for later analysis. Both the materials in contact with the sample and the sample storage time can affect the chemical composition and therefore the measurement results. Specifications for sampling equipment and for calibration, conditioning, cleaning, and re-use procedures are described in the European Standard (European Committee for Standardization, 2001), which also lists the following materials as appropriate for those parts of the equipment that will be in contact with the odor sample: stainless steel, glass, polytetrafluoroethylene (PTFE), tetrafluoroethylene-hexafluoropropylene copolymer (FEP), and polyethyleneterephalate (PET, Nalophan). Polyvinylfluoride (PVF, Tedlar), FEP, and PET are appropriate materials for making sample containers. Tedlar bags (10-50 L) that are inflated in the field using portable wind tunnels or negatively pressurized canisters are commonly used.

Odor Concentration Measurements

Methods for measuring odor concentrations include sensory methods, analytical methods, and "electronic noses" for specific odorous gases. An electronic nose is an array of gas sensors that is combined with pattern recognition software to mimic human olfactory response (Lacey, 1998). Sensory methods include olfactometry, scaling, and rating, of which olfactometry is the most widely used. Olfactometry involves collecting and presenting odor samples (diluted or undiluted) to selected and screened panelists under controlled conditions using scentometers and dynamic olfactometers. Scentometers are portable field measurement instruments that can be used for direct sampling and measurement of ambient air and have been used as the basis for setting property line odor concentration standards by several states (e.g., Colorado, Missouri, Montana, North Dakota, and Kentucky) and various cities. Scentometers have also been used for field odor measurement at numerous livestock and poultry operations in the United States (Miner and Stroh, 1976; Sweeten et al., 1977, 1983, 1991) and in data collection for nuisance litigation (Sweeten and Miner, 1993). Dynamic Triangle Forced-Choice Olfactometers, which offer more accuracy in odor measurement than scentometers, are the instrumentation of the choice for the ASTM (American Society for Testing and Materials) and European Standards and are widely used for odor research. Guidelines for the design, construction, calibration, and operation of olfactometers are given in the European Standard (European Committee for Standardization, 2001). Specific requirements for panel size and selection with respect to behavior, variability, and sensitivity of panel members are also provided. The minimum panel size in any measurement is four, but larger numbers are recommended to improve repeatability and accuracy. The scentometer may be more appropriate for ambient measurements (property line, downwind of source, etc.) than the olfactometer.

Instruments available to identify and measure the concentrations of specific odorants include gas chromatography coupled with mass spectrometry (GC-MS) for component identification. Some of these methods are sensitive in detecting compounds at very low concentrations. Peters and Blackwood (1977) reported difficulty in positively identifying compounds present in feedlot air samples using GC-FID. (Low peak values precluded the use of GC-MS for amines.) As a result of the low concentrations of many AFO odorants, their components may have to be concentrated prior to analysis using methods such as solvent desorption, thermal adsorption (Zahn et al., 1997), or solid-phase microextraction (SPME) (Zhang et al., 1994). It must be emphasized that chemical techniques should be buttressed by sensory methods to correlate instrumental results with human odor perception.

The relatively high cost per sample of odor panels has created the need for reproducible, inexpensive instruments (electronic noses) capable of making measurements that correlate with the human olfactory response (Lacey, 1998). Lacey (1998) and Mackay-Sim (1992) listed several electronic approaches to volatile gas (odor) detection, including metal oxide semiconductors, field-effect transistors, optical fibers, semiconducting polymers, and piezoelectric quartz crystal devices. These approaches raise the possibility of remote odor monitoring or surveillance networks for individual compounds or odorant mixtures. Piezoelectric crystals are sensitive to changes in surface mass caused by adsorption of gas molecules. As mass is added to the surface, the resonant frequency decreases and can be measured precisely. The crystal surface can be made to respond to single chemicals or groups of chemicals. Some sensors may be affected by water vapor, methane, and temperature (Lacey, 1998).

Electronic methods should be tested against olfactometry results to be validated against human sensory responses.

FINDING 5. Standardized methodology for odor measurement have not been adopted in the United States.

RECOMMENDATIONS:

- **Standardized methodology should be developed in the United States**

for objective measurement techniques to correspond to subjective human response.

- **A standardized unit of measurement of odor concentration should be adopted in the United States.**

MEASURING EMISSIONS

The method selected for measuring emissions will depend on the type of emission and whether it is from a point source (e.g., an exhaust vent from a mechanically ventilated building; Figure 4-2) or an area source (e.g., a waste lagoon; Figure 4-3).

The emission rates for low-level point sources (LLPSs; Figure 4-2) may be determined by measuring concentrations (mass per unit volume) and volumetric flow rates (volume per unit time) at the emitting points and multiplying the two measurements. The emission rates will be expressed as mass per unit time. An alternative procedure consists of measuring the ambient concentrations upwind and downwind (off-property) and back-calculating the emission rate using dispersion modeling. The Air Pollution Regulatory Process (APRP) addresses off-property impacts on the public. For criteria pollutants, EPA regulations stipulate that the 24-hour downwind concentration should not exceed the NAAQS at the property line or at the nearest occupied residence.

The emission rate for a ground-level area source (GLAS; Figure 4-3) may be determined using "flux chambers" or micrometeorological techniques or by measuring upwind and downwind concentrations and back-calculation of flux using dispersion modeling. The units of flux will be mass per unit area per unit time.

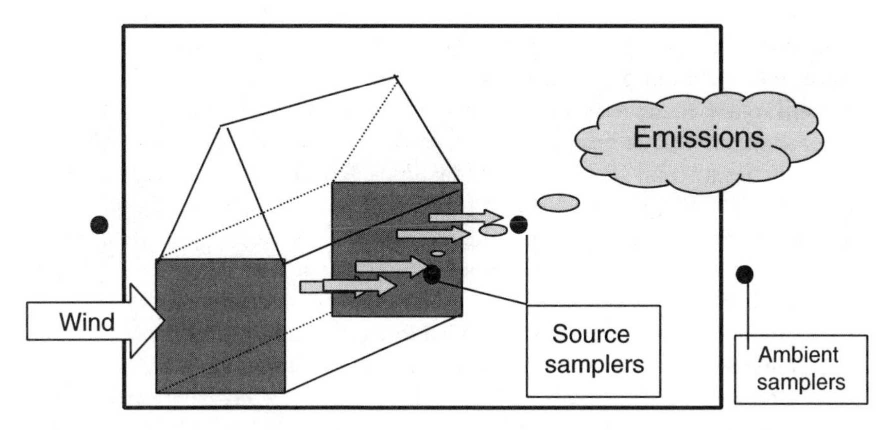

FIGURE 4-2 Schematic illustrating the essential elements associated with measurement of emissions from agricultural sources that can be characterized as low-level point sources such as cotton gins, feed mills, grain elevators, and oil mills.

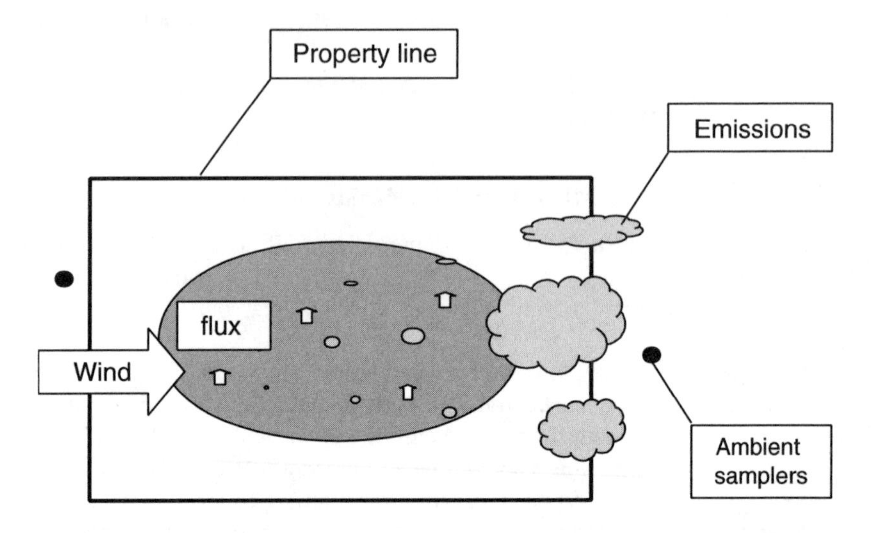

FIGURE 4-3 Schematic illustrating the essential elements associated with measurement of emissions from agricultural sources that can be characterized as ground-level area sources such as dairies, cattle feed yards, field operations, and agricultural burning.

Point Sources

In theory, measurement of the emission rates of gaseous substances from a mechanically ventilated animal facility requires only the concentration of the substance being emitted and the ventilation rate, but accurate measurement of these two factors is difficult in practice. Ventilation rate is affected by many factors including the length of time the fans operate, fan design, fan speed, fan maintenance, motor startup time, static pressure, outside wind speed, wind direction, and infiltration. In practice, measurement of the concentration of an emitted substance is often difficult because of frequently changing ventilation rates. For example, in negative-pressure ventilated facilities, fans do not usually operate continuously, but rather cycle on and off for short periods of time. These short bursts of ventilation (followed by little or no ventilation) are necessary to create sufficient negative pressure to bring in air through the inlets at a proper speed to promote air mixing. If fans operated continuously, animals might become chilled or excess fuel would have to be expended to warm the buildings. Concentration measurement often becomes more difficult at high ventilation rates because substances may be diluted and be present only in very low concentrations. Since ventilation rates can be very high during warm weather and/or with large animals that give off large quantities of heat, even small absolute errors in the measurement of the concentration of an emitted substance can result in significant errors in emission rates. Errors of 25 to 100 percent in ammonia emission rates have been common in the past.

Mechanically Ventilated Buildings

Mechanically ventilated confined animal facilities (Figure 4-4) may be ventilated using positive-pressure, negative-pressure, or neutral ventilation systems. Positive-pressure systems are equipped with fans that force fresh air into the building and thus create a slight positive static pressure; fresh air is usually circulated within the building by mixing fans. Negative-pressure systems exhaust air from the building and thus create a negative pressure (typically between 0.05 and 0.10 inch of water column) within the building; fresh air enters the building and is mixed through carefully placed air inlets. Neutral-pressure ventilation systems match fans forcing air into the house to fans exhausting air out. Pressure differentials in these types of facilities are essentially zero.

Mechanically ventilated facilities are engineered to provide specific air exchange rates related to the needs of the animal (low air exchange rates for young animals in cold weather, high rates for large animals in hot weather). In the case of poultry, typical ventilation rates range from 0.1 to 10 cubic feet per minute

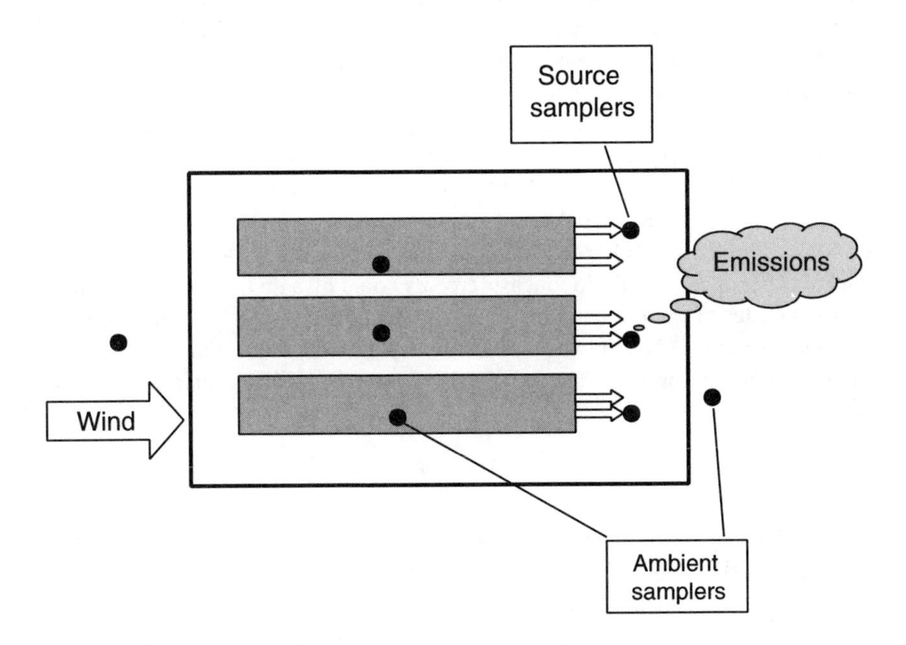

FIGURE 4-4 Schematic illustrating the essential elements associated with the regulation of emissions from agricultural sources that can be characterized as low-level point sources such as tunnel-ventilated AFOs.

(cfm) of air exchange per animal (MWPS, 1990; Lacy and Czarick, 1992). For swine, these rates range from 2 to 300 cfm per animal. Recommended air exchange rates for dairy cows in confined housing range from 50 to 1000 cfm per animal (MWPS, 1990). Ventilation rates in mechanically ventilated facilities depend on the number of fans and the length of time they operate. Relatively precise control is provided by timers, thermostats, and in many cases, computer equipment. Ventilation rates differ and/or change depending upon house design, animal density, animal age, climate, season, feeding program, and so forth. Again, very low ventilation rates are typical when animals are young and heat production by the animals is minimal; high rates are necessary for older animals when heat produced by the animals becomes a concern and must be removed. Measurement of low ventilation rates (e.g., during cold weather) is difficult and prone to error.

Naturally Ventilated Buildings

Naturally ventilated facilities (Figure 4-5) rely on wind currents to provide fresh air needs to the animals and to remove excess moisture, ammonia, CO_2, et cetera. In most cases, air enters these naturally ventilated facilities through openings in the sidewalls. These openings are typically fitted with adjustable curtains or panels that provide rudimentary environmental control (MWPS, 1989).

Ventilation rates in naturally ventilated houses depend on wind velocities and adjustment of the curtain openings. Measuring ventilation rates accurately in naturally ventilated houses is extremely difficult due to the dynamics of outside weather and wind conditions.

The emission rate for an LLPS (Figure 4.4 and 4-5) will be expressed as mass per unit time. The APRP could include measurements of “ambient” concentrations using TSP, PM10, or PM2.5 samplers upwind and downwind from the source off-property. Emission measurements could also include source sampling emission rates at the point source or inside the house using the assumption that the concentrations of pollutants emitted are equal to the measured concentrations of the indoor environment. The APRP addresses off-property impacts on the public.

Area Sources

To determine emissions from area sources, it is necessary to take into account the local meteorology and the wind field. Gases and aerosols are exchanged between the earth’s surface and the atmosphere through turbulent processes that take place near the surface. Dispersion by atmospheric turbulence is orders of magnitude greater than molecular diffusion, and tracer gases often provide a useful measure of wind diffusion. Vertical emissions through a horizontal plane par-

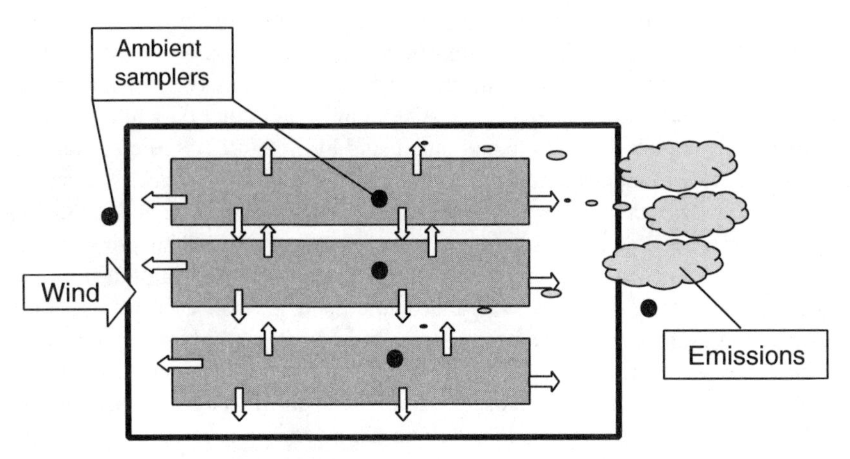

FIGURE 4-5 Schematic illustrating the essential elements associated with the regulation of emissions from agricultural sources that can be characterized as low-level point sources such as naturally ventilated AFOs.

allel to the ground result from vertical turbulent diffusion processes (Lamb et al., 1985).

Emissions of gases or aerosols from area sources are often expressed in terms of fluxes, or mass emission rates per unit area perpendicular to the direction of the flux. Fluxes are measured at a variety of scales, including small, surface-layer, and mixed-layer scales (for a discussion of small and surface-layer scales, see Fowler [1999]). In all of the measurement methods there exist considerable uncertainties, as well as advantages and disadvantages. At the small scale (e.g., in a tent on a field or a chamber over a water surface), the environment is enclosed and thus the surface being studied is altered. At the surface- and mixed-layer scales, fluxes are determined by measuring the vertical turbulent transport. The height of the measurement dictates the area of the surface over which the measurement is averaged and, thus, the scale of the measurement.

Small Scale (10^{-1} to 10^{2} m^2)

A number of enclosure techniques exist to estimate fluxes at the small scale; some involve chambers and wind tunnels. In an enclosed-chamber experiment, a chamber is placed either on or around the source. The experiment can be operated in a flow-through manner, in which the airflow rate is known and the concentrations into and out of the chamber are measured. It is also possible to conduct an enclosure experiment in a dynamic, nonflow manner, in which the change in the chamber concentration of the substance of interest in the chamber is monitored over time. Conceptually, the enclosure experiment is relatively straightforward.

Its advantages include the ability to conduct process-level or sensitivity studies of the factors that control emissions. Significantly less infrastructure may be needed than that required for the micrometeorological techniques used in larger-scale studies. The major drawback of enclosure techniques is that they alter the environment of the emission source and thus could bias the measured emission rate. They also give a measurement in a very limited region of space, so that it is difficult to both capture the spatial variability of the emission rate and to integrate over larger areas. For example, if there are a few "hot spots" where the surface of a lagoon is disturbed, measurement at the hot spots will be too high, measurements elsewhere will be too low, and neither will give an accurate picture of average emission rates. If the trace species of interest is highly reactive or water soluble, losses to equipment walls may complicate the experiment. Nevertheless, enclosure techniques have been used to study trace gas fluxes from soils, lagoons, and vegetation. They can be especially useful in determining the relative emission rates of gases.

Surface-Layer Scale (10^2 to 10^6 m^2)

At the surface-layer scale, micrometeorological techniques or mass balance methods can be used to measure area source fluxes. Micrometeorological methods include eddy correlation, eddy accumulation and other conditional sampling techniques, and gradient and difference methods. These techniques are typically implemented by installing instrumentation on tower platforms, thus requiring substantial experimental infrastructure. Disjunct eddy covariance is a new technique (Rinne et al., 2001) that allows measurement of trace gas fluxes with relatively long time intervals between quick gas samples, rather than continuous sampling.

A basic requirement for all of the micrometeorological methods to be successful is a horizontally homogeneous surface (long, uniform fetch). One must also contend with flow distortion caused by the tower itself, as well as by sensors installed on the tower. Micrometeorological methods have been widely used to measure CO_2, water vapor, and biogenic hydrocarbon emissions from forest or agricultural canopies. The requirements of uniformity of the canopy and fetch would likely not be easily met for animal confinement operations, but may be applicable for measuring emissions from some land applications or over slurry lagoons. Some, but not all of the techniques require fast-response chemical sensors. These techniques also require determination of micrometeorological parameters including eddy diffusivities for momentum, heat, or water vapor; latent and sensible heat fluxes; and atmospheric stability correction factors. Limitations in micrometeorological sensors are also of concern (Lapitan et al., 1999). Spatial variability due to hills, trees, buildings, varying soil fertility, and biomass density complicates the use of micrometeorological methods for flux determinations.

Eddy Correlation or Eddy Covariance

This is the most direct micrometeorological method of determining vertical fluxes; however it is difficult to implement. In this technique, the flux is determined by measurement of the covariance of the fluctuation of the concentration of the species of interest with the fluctuation in the vertical wind speed. Thus, the method requires concurrent, contiguous measurements of the species concentration and the vertical component of the wind, both sampled at high frequency (>1-10 Hz). In addition, it is important to obtain good vertical alignment of the vertical wind component sensor, and flow distortion around other sensors or the sampling tower itself can cause problems. Because of the fast sensor requirements, eddy correlation has been used to measure fluxes of carbon monoxide (CO), CO_2, water vapor, O_3, NO_x, N_2O, SO_2, and CH_4, but not many other trace gas species (Guenther and Hills, 1998).

Eddy Accumulation

This method is a variation of the eddy correlation technique which relaxes the requirement for fast chemical sensors by "conditional sampling," in which samples are collected in two or more containers based on the vertical wind velocity component. The sample collection rate is proportional to the vertical wind speed. In this way, samples can be collected and analyzed later. The trade-off is that the method requires fast, accurate flow control with a good dynamic range, rather than fast chemical sensors. In eddy accumulation, the concentrations in the sample reservoirs are typically not very different, so detecting statistically significant differences between the reservoirs may be difficult. Also, any mean offset in the vertical component of the wind must be removed in real time.

Relaxed Eddy Accumulation (REA)

This is another conditional sampling technique that involves sampling into two reservoirs based on the direction of the vertical wind velocity component. In REA, the requirement for proportional sampling is relaxed, permitting constant sampling rates and thus greatly simplifying the technique. The disjunct eddy accumulation technique further simplifies the operation. REA includes an empirical constant b, which must be known a priori (Gao, 1995). The value of b has been determined experimentally to be approximately 0.6. Like the eddy accumulation method, REA depends on the precision of the measurement method in order to measure a statistically significant difference between two samples.

Flux Gradient and Difference Methods

These commonly used methods do not require fast-response sensors. The flux is estimated from concentration differences between two or more levels and from the wind speed profile. The "constant" of proportionality (eddy diffusivity) must be characterized, either by measuring the energy balance, by measuring the vertical wind profile, or by measuring the concentration gradient along with the flux using the eddy correlation method for another trace gas. Uncertainty in the determination of eddy diffusivity is a major source of uncertainty in the flux gradient methods (Lapitan et al., 1999). Like other micrometeorological methods, flux gradient techniques are subject to uncertainties associated with spatial and temporal heterogeneities.

Bowen Ratio

This is an indirect technique that is based upon the surface energy budget. The Bowen ratio is the ratio of the sensible heat to the latent heat. This method does not require eddy flux measurements or stability corrections. It does require measurement of the incoming net radiation at the earth's surface and the soil heat flux, as well as the concentration difference for the species of interest, along with concurrent flux and difference measurements of temperature, humidity, or other trace species. The Bowen ratio method fails under conditions of low energy availability such as during the night or during periods of precipitation. Like all of the micrometeorological methods, the Bowen ratio technique is subject to errors associated with the chemical and micrometeorological sensors and the spatial and temporal variability of fluxes within the source area of interest.

The Box Model

As opposed to the above micrometeorological techniques, the box model method works over smaller areas of fetch and so may be more applicable for some AFOs than the micrometeorological techniques discussed above. The method is a direct technique requiring no empirical relationships. In it, an imaginary box is constructed around some measurement volume so that the transfer of mass into and out of the walls of the box is measured (Shaw et al., 1998). In the most common configuration for flux measurements, the vertical profile of both the horizontal wind speed and the species concentration must be sampled through the entire downwind plume. The height of the plume, and thus the height to which sampling must be conducted, will depend on the atmospheric stability. The mass balance method assumes steady-state conditions, homogeneous horizontal winds, no other internal sources or sinks of the species of interest, and zero or known background concentrations. The box method is not suitable for highly reactive gases such as ammonia that may be rapidly deposited on surfaces. Other problems include disturbing the soil or vegetation to be measured (such as increasing the temperature) and thereby perturbing the rate of emissions. Also where turbulence or fast photochemistry strongly influence emissions or ambient concentrations, box methods must be used with caution.

Dispersion Modeling

In this method, upwind and downwind concentration measurements are made, and the emission rate is back-calculated from a dispersion simulation, usually based on a Gaussian dispersion algorithm, making assumptions regarding the characterization of the atmospheric stability and dispersion. This method can be applied to area, line, and point sources and would be applicable to a number of animal feeding operations. Due to the many assumptions required, Gaussian dispersion models can be assumed to be good to no better than 50 percent, with sources of uncertainly also introduced by chemical sensors, as well as stability classification.

Atmospheric Tracers

Atmospheric tracers, such as sulfur hexafluoride (SF_6), can be used either to measure emission rates "directly" or to improve the dispersion characterization in a dispersion model, thus decreasing the uncertainty in that technique. Under certain conditions, direct flux measurements can be made if the tracer release can adequately simulate the emission source (e.g., point releases to mimic a smokestack emission; line source releases to mimic a heavily traveled road) and it can be shown that the tracer disperses in the same manner as the emitted species. In this case, the emission rate can be calculated from the known emission rate of the tracer, multiplied by the ratio of the gradients of the emitted species and the tracer gas. Another application of an atmospheric tracer relevant for AFOs is in the determination of air exchange rates inside enclosures (Lapitan et al., 1999) such as buildings.

Mixed-Layer Scale (1 to 1000 km^2)

At the mixed-layer scale, micrometeorological and mass balance techniques can be utilized, with the sample platform now installed on an aircraft or balloon. This allows for spatial averaging over larger areas but introduces additional expense and infrastructure requirements, along with new difficulties, some associated with inlet design for aerosol samplers or sensors, and others with inhomogeneities in surface fluxes at the scale of the flight lengths utilized.

FINDING 6. The complexities of various kinds of air emissions and the temporal and spatial scales of their distribution make direct measurement at the individual farm level impractical other than in a research setting. Research into the application of advanced three-dimensional modeling techniques accounting for transport over complex terrain under thermodynamically stable and unstable planetary boundary layer (PBL) conditions offers good possibilities for improving emissions estimates from AFOs.

RECOMMENDATION: EPA should develop and carry out one or more intensive field campaigns to evaluate the extent to which ambient atmospheric concentrations of the various species of interest are consistent with estimated emissions and to understand how transport and

chemical dynamics shape the local and regional distribution of these species.

AIR QUALITY MONITORING

Air quality monitoring involves measuring concentrations in the ambient air. Currently, almost all monitoring is carried out in cities; little is done in rural areas where AFOs are found. Considerations involved in establishing a monitoring program include the purposes of monitoring; the emissions of interest; analysis methods, and the precision and accuracy needed; the use of mobile versus stationary monitors; the number and locations of sampling sites; sampling frequency and averaging times; the intrusiveness of monitors; and costs.

For example, to assess the magnitude of potential health and environmental impacts of emissions from AFOs, it would be helpful to have much more data on ambient concentrations than are now available, starting perhaps with mobile monitors for substances expected to have the greatest health and environmental effects, in areas with high concentrations of livestock, and at times when emissions are expected to be greatest and meteorological conditions are not conducive to rapid vertical or horizontal dispersion (a stable atmosphere and low wind speeds).

FINDING 7. Scientifically sound and practical protocols for measuring air concentrations, emission rates, and fates are needed for the various elements (nitrogen, carbon, sulfur), compounds (e.g., NH_3, CH_4, H_2S), and particulate matter.

RECOMMENDATIONS:

- **Reliable and accurate calibration standards should be developed, particularly for ammonia.**
- **Standardized sampling and compositional analysis techniques should be provided for PM, odor, and their individual components.**
- **The accuracy and precision of analytical techniques for ammonia and odor should be determined, including intercomparisons on controlled (i.e., synthetic) and ambient air.**

SUMMARY

Assessment of the health and environmental effects associated with air emissions from animal feeding operations requires a substantial increase in both the accuracy of estimates of emissions of substances of interest and the accuracy of measurements of their concentrations. Concentrations are important for the determination of exposure and emission rates.

Concentration measurement requirements include real-time capability, adequate precision and accuracy, and the availability of suitable calibration standards. Not all of the air emissions of concern from AFOs can be measured by techniques that currently meet these requirements; for example, methods for measuring PM suffer from lack of good calibration standards.

Although measurements of air emission rates have been reported for numerous substances and from a variety of source types (point, line, and area), there are relatively few reports of emission measurements from operations within AFOs, for a variety of reasons. The paucity of emission measurements from AFOs is likely due at least in part to a lack of resources available for this research area. The availability of concentration measurement methods is a prerequisite for emission rate determination. Many of the emission rate methods used for other sources could be adapted to determine emission rates of substances from AFOs. Given the variability of AFOs in such matters as configurations, animal populations, climate, and management practices, the variability of emission rates is expected to be great temporally, spatially, and from one AFO to another.

5

Approaches for Estimating Emissions

INTRODUCTION

In the absence of effective and efficient means to measure air emissions from each animal feeding operation (AFO) directly, regulatory or management agencies seeking to mitigate emissions need reasonably accurate means to estimate them and attribute them to particular operations or activities. The committee has considered three possible approaches: (1) emission factors from representative AFOs; (2) regression analysis equations that relate air emissions to specific aspects of individual AFOs; and (3) process-based models that estimate the flows of emission-generating substances through the sequential processes of the farm enterprise. The committee favors the third approach which is discussed in greater detail in this chapter.

The "emission factors" approach is based on measurement of air emissions from a defined set of "average" AFOs that presumably represent a substantial proportion of the national population. Emission factors, expressed as the mass of each substance emitted per animal or other base unit per unit of time, are then used to estimate air emissions from other AFOs, which are assumed to fall into one of the categories in the defined set of average AFOs used to estimate the emission factors. Weaknesses in this approach noted in the committee's interim report (NRC, 2002a) include the difficulty of defining a small set of average AFOs that can represent the broad range of AFOs for varied livestock industries in various geographic regions. As with the other approaches, obtaining the data upon which to base useful estimates of air emissions requires many direct measurements, which do not exist today in sufficient numbers to provide reasonable confidence in the results.

The "regression analysis" approach uses standard least-squares multivariate regression equations to relate measures of air emissions to various factors that are hypothesized to affect them (e.g., number of animals, animal type and production system, productivity, housing, manure management, weather, climate). Once the equations are verified, they are used to gauge the importance of various factors that determine emissions and to estimate air emissions from other AFOs based on their individual characteristics. The weakness of this approach is the assumption that the within-factor variation of emissions is small relative to the among-factors variation for any category of operation. In addition, the current database for estimating regression equations is very limited and a major effort would be required to obtain them.

The "process-based" approach follows the fate of relevant elements (e.g., nitrogen, carbon and sulfur) step by step through the animal feeding process and identifies the chemical transformations that take place. It provides estimates of the characteristics and amount of air emissions that occur at each step as controlled by a mass balance approach (i.e., the emission of an element from the system, or from a part of the system, is equal to the input of that substance minus any accumulation that might occur). The advantages and limitations of this approach are described in some detail later in this chapter.

EMISSION FACTOR ESTIMATES

The goal of the U.S. Environmental Protection Agency (EPA, 2001a) is "to develop a method for estimating [air] emissions at the individual farm level that reflects the different animal production methods that are commonly used at commercial scale operations." The approach was intended to provide estimates of total annual air emissions from AFOs for defined geographic areas by kind of animal and manure handling practices for each of eight kinds of emissions. It did this with a model farm construct that provides estimates of average annual emissions per EPA animal unit (AU) for twenty-three model farms (two for beef, eight for dairy, two for poultry-broilers, two for poultry-layers, two for poultry-turkeys, five for swine, and two for veal; EPA, 2001a). Each model is defined by three variable elements that describe manure management practices for typical large AFOs: (1) confinement and manure collection system, (2) manure management system, and (3) land application. The manure management system is further subdivided into solids separation and manure storage activities. Insofar as combinations of these elements are regionally distinctive, the model farms also reflect regional variations in air emissions.

Model farms, as described by EPA (2001a), are useful for aggregating emission rates across diverse sets of AFOs. A model farm can be used to represent the *average* emissions across some geographic area over some period of time per unit capacity of a class of farms (e.g., all pig farms in the United States that use an enclosed house with pit recharge and irrigation of supernatant onto forage land; model farm S2).

The utility of this kind of model farm construct depends on the following:

- defining models in which the dependent variable—the amount of each air emission per unit of time—is closely related to independent variables that accurately depict real feeding operations and that explain a substantial share of the variation in the dependent variable;
- providing accurate estimates of the relationship between the dependent and independent variables; and
- having estimates of the relationships between dependent and independent variables that clearly distinguish among the kinds of AFOs being modeled.

A critical requirement for estimating the appropriate emission factors is a statistically representative survey of emissions from a class of AFOs over several iterations of the time period to be represented. The size of the sample required to estimate the mean emission rate with a given degree of accuracy increases with the variability in the dependent variable to be measured (e.g., the average emission rate) across the set of independent variables that affect it. Independent variables that have been discussed include animal type and age, diet, local climate, building type, land application method, and management practices. To the extent that some of these variables change over time (e.g., trends in farm organization, location, practices, and technology), updating of estimates and estimates of trends may be required.

The model farm construct is represented by Equation 5-1:

$$E = \Sigma_i \, (w_i \bullet e_i) \qquad \text{(Eq. 5-1)}$$

in which the total emission (E) of a particular pollutant from an AFO during a period of time is the product of the emission (e_i) from each unit (i) on the model farm and the number of units (w_i) of that type, summed over the farm.

One use of model farms might be to predict emission rates and local effects of a single AFO or a cluster of AFOs in a small area. This use differs from that described by EPA (2001a), and it would require a detailed model or models describing the effects of selected variables on the rates of emissions and their downwind concentrations. An example is an odor dispersion model that predicts odor intensity as a function of time at various locations, given information on odor sources and local meteorological conditions. More data (perhaps hourly) and statistical analyses of the relationships between various explanatory variables and pollutant concentrations or impacts are required.

The committee believes that EPA's proposal (2001a) is inadequate to meet these standards. It does not provide a method to adequately determine air emissions from individual AFOs because both the model farm construct and the data are inadequate. The model farm construct used by EPA (2001a) cannot be supported for estimating air emissions from an individual AFO because it cannot account for a great deal of variability among AFOs. In particular, important factors not included in the EPA model include animal feeding and management practices; animal productivity; housing, including ventilation rate and confinement area; use of abatement strategies such as sprinklers to decrease dust; and physical characteristics of the site such as soil type and whether the facility is roofed. In addition, emissions are likely to differ for different climatic (long-term) and weather (short-term) conditions including temperature, wind, and humidity. Accurately predicting emissions on individual AFOs would require determination of emission factors that reflect these characteristics.

More specifically, improvements in the model farm construct would be needed for both discrete variables (e.g., management, confinement conditions, location) and continuous variables (e.g., nutrient input, productivity, meteorology). Concerns about the quality of data (much of it not reviewed), great gaps in the data, inappropriate use of data, and representativeness of the data were discussed in the committee's interim report (NRC, 2002a; see Appendix L).

FINDING 8. Estimating air emissions from AFOs by multiplying the number of animal units by existing emission factors is not appropriate for most substances.

RECOMMENDATION: The science for estimating air emissions from individual AFOs should be strengthened to provide a broadly recognized and acceptable basis for regulations and management programs aimed at mitigating the effects of air emissions.

While the committee favors a process-based modeling approach, there are some substances, such as particulate matter (PM) and odor for which an emission factor approach may still be the best approach. An example of an emission factor approach utilizing mass balance constraints for the emission of PM from a feed mill is given in Appendix I. A critical need exists to determine emission factors or fluxes for AFO emissions of PM and odor-causing constituents emanating from either area or point sources. Specific short-term research needs include the following:

- development of protocols to determine emission fluxes from downwind concentration measurements,
- evaluation of dispersion models to determine the best method to calculate fluxes,
- determination of the best method to establish volumetric flow rates for mechanically and naturally ventilated AFOs,
- comparison of emission factors for naturally ventilated and mechanically ventilated AFOs, and

- determination of the applicability of utilizing PM samplers designed and calibrated for the urban environment in a rural setting.

MULTIPLE REGRESSION APPROACH TO DEVELOP EMISSION FACTORS

Given the inadequacy of the model farm construct described above, the committee discussed developing a different construct by systematically identifying factors that contribute to changes in emission rates. The approach would still be empirically based on measured total emissions from individual representative farms. A comprehensive and representative data set would be developed largely from new investigations. The various emission measurements (e.g., for ammonia [NH_3] and methane [CH_4]) would be regressed against several independent variables hypothesized to affect emission rates (e.g., animal type, management system, abatement strategy). Significant terms could be included in an empirical model to predict emissions from farms. Principal component analysis might be used to determine which factors contribute the most to emission rates and to evaluate whether certain practices give significantly different results than others. Compared to the existing emission factor approach, this alternative would result in more accurate estimates of emissions from individual farms and the identification of strategies to decrease emissions.

A comprehensive data set would have to be developed from existing literature and new investigations. Given the paucity of research in which total emissions from individual farms have been measured accurately, most of these data would have to come from new studies. The measurements would have to be accurate and representative of the emissions that occur for the time period of interest (e.g., an hour, or a full year). The data set would also have to include farms representing all factors that are hypothesized to affect emissions. In particular, it should include enough distinctive measurements to differentiate the types of classifications made in the EPA model farm construct (e.g., animal type and manure management systems), as well as additional factors not included in the EPA model that are likely to affect emissions. For example, farms would have to be included that represent different types of animal management; levels of animal productivity; housing including ventilation and confinement area; use of abatement strategies such as sprinklers to decrease dust; and physical characteristics of the site, such as soil type and whether the facility is roofed. The data set would also have to be replicated under different climatic and weather conditions and would require adequate replication on farms of similar type so that the variance not accounted for in measurements could be determined. Developing a robust set of equations would require sampling hundreds of AFOs representing different management and meteorological conditions. The cost of accurately measuring emissions on the number of AFOs (i.e., thousands) that would be needed to replicate all common situations would be very high, and the time required would probably be too long in the face of the demands for regulation.

PROCESS-BASED MODELING APPROACH

The committee recommends using a process-based modeling approach to predict emissions from both individual AFOs and regions. A process-based approach would involve analysis of the farm system through study of its component parts. It would integrate mathematical modeling and experimental data to simulate conversion and transfer of reactants and products through the farm enterprise (Denmead, 1997; Jarvis, 1997). In many cases, this approach makes use of mass balance equations to represent mass flows and emissions of major elements (e.g., nitrogen, carbon, sulfur) through the system. In other cases (e.g., particulate matter and odors), alternative approaches such as emission factors may be the best way to predict emissions. In either case, models are needed to estimate emissions from individual AFOs to identify farms that should be targeted for control strategies and to predict the impact of management techniques to decrease emissions. Models are also needed to estimate emissions across specific regions and globally. Currently, emission factors are multiplied by the animal inventories to make these estimates. Process-based models may be used with data on animal and crop inventories and specific information on management strategies. Additional data may have to be collected by the agricultural census to use new process-based models.

FINDING 9. Use of process-based modeling will help provide scientifically sound estimates of air emissions from AFOs for use in regulatory and management programs.

RECOMMENDATIONS:

- **EPA and the U.S. Department of Agriculture (USDA) should use process-based mathematical models with mass balance constraints for nitrogen-containing compounds, methane, and hydrogen sulfide to identify, estimate, and guide management changes that decrease emissions for regulatory and management programs.**
- **EPA and USDA should investigate the potential use of a process-based model to estimate mass emissions of odorous compounds and potential management strategies to decrease their impacts.**
- **EPA and USDA should commit resources and adapt current or adopt new programs to fill identified gaps in research to improve mathematical process-based models to increase the accuracy and simplicity of measuring and predicting emissions from AFOs (see Chapter 7).**

Components of the AFO system

The system for animal production includes several components shown in Figure 5-1. Of course, all AFOs by definition contain animals on a lot or in housing. Emissions can occur directly from living animals or from animal mortality that is discarded. The animals also produce manure that is either handled on the farm or exported to other farms. In either case, manure is applied to soil on which crops are grown. Crops that produce feed for animals may or may not have been fertilized with manure, but invariably they will have been fertilized in some way. Emissions can occur from many components of the AFO. For example, ammonia may be volatilized from the animal housing, manure storage, or field. The amount lost from one component of the farm affects the amount that can be emitted from subsequent components. For example, if nitrogen is volatilized from an animal house, it is not available to be volatilized again from manure storage or field.

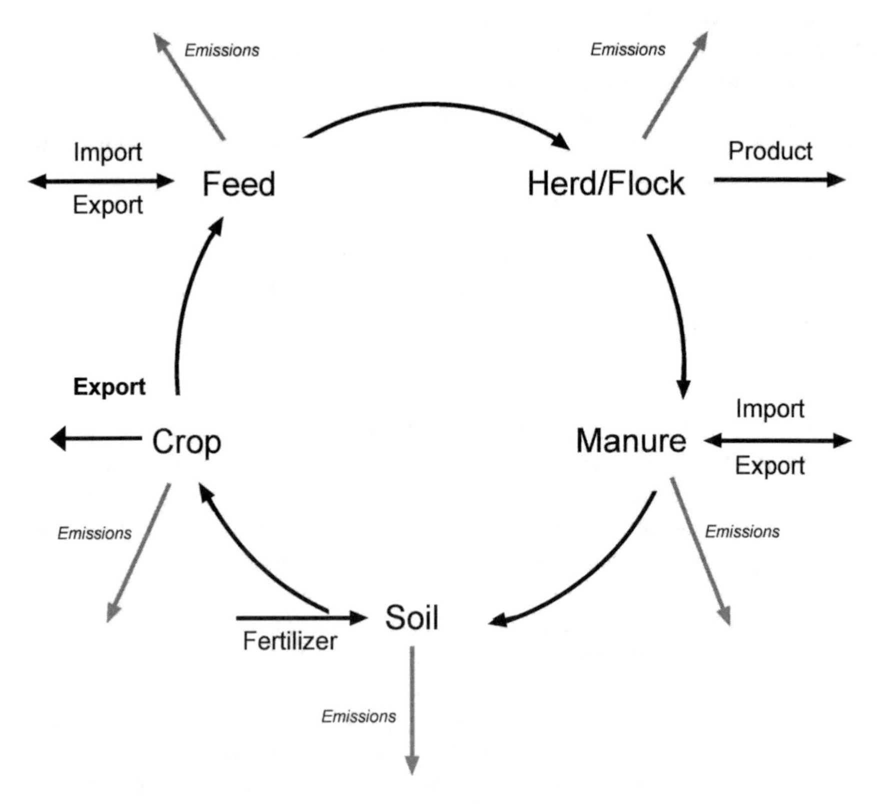

FIGURE 5-1 A schematic representation of a process-based model of emissions from an animal production system. Emissions can be to air or water.

Moreover, transformations that occur in one of the farm components might affect emissions and further transformations in other components. The committee recommends modeling the major farm processes in such a way as to include the effects of transformations in each farm component on the others.

An animal feeding operation may not include all of the components shown in Figure 5-1 on one farm. However, a complete system for animal production must include crop production and manure utilization or disposal even if these activities are carried out on separate farms. When emissions from AFOs are considered, the environmental impact depends on management of the complete production system. For example, an AFO may decrease emissions from its property by importing feeds or exporting manure to a separate farm. Unless the handling of crops or manure is superior on the other farms, the overall emissions to the environment would not be decreased; they would simply be moved off farm and possibly increased. This raises an important issue related to regulation of emissions from individual farms. If crop production and manure handling are not regulated in a similar way on AFO and non-AFO farms, there will be an incentive to outsource crop production and manure use to decrease AFO emissions, but with no benefit to the environment. Emissions from the operations that produce feeds for animals, including crop farms and feed processors, must be considered as part of the animal production system. In addition, the way in which different foods are processed and used once they leave the farm affects total air and water emissions.

Comparison of the Committee's Approach to the EPA Model Farm Approach

The EPA model farm approach (EPA, 2001a) uses emissions from housing, manure storage, and field application and adds them together. Using this approach, one would predict that a technology to decrease emissions from manure storage (e.g., covering manure lagoons) would decrease total farm ammonia emissions by the amount that was prevented from leaving the lagoons. In reality, this ammonia would be concentrated in the lagoon liquid—increasing the emissions in the barn when flushing with lagoon liquid and in the field during land application. In a process-based approach, rather than simply adding the emissions from each farm element, a mathematical model is used to represent the interactions between the system components. This alternative to EPA's model farm approach would resolve many problems such as the one cited above and prevent predicting more manure nitrogen volatilized than excreted. Manure nitrogen production is used to predict the nitrogen available for conversion to ammonia for possible subsequent volatilization.

Development of a process-based model does not obviate the need for data collection, but it enables the use of data representing only part of the farm system, and will help identify gaps in the existing literature. In fact, data will be needed to represent transformations that occur on farms other than air emissions (e.g., organic N degradation in lagoons). Some of the necessary data may be available in the existing literature; however, additional studies will be necessary as outlined for specific emissions in the sections that follow. The accuracy of process-based models will still need to be determined with collected research data. The accuracy to which model inputs can be determined and their impacts on the model estimates will also have to be determined using sensitivity analysis.

Predicting Differences Among Farms

The process-based model has to account for differences among farms that may affect substances emitted, emission rates, pollutant concentrations, or flow of mass between farm components. These differences among farms include animal type and species; farm size (animal number and area) and density; animal or crop production system and productivity; housing type; manure collection, storage, and application; soil type and capacity; cropping systems (including tillage method, crop rotation, and irrigation practices); and weather (short term) and climate (long term). The way in which each of these differences among farms should be represented in the model is discussed in a general way below, and more specifically for individual emissions in the sections that follow.

The size of a farm, as represented by the number of animals or area of cropland, is likely to affect the amount of emissions from it. For any animal type (e.g., swine, dairy cattle, chickens), the rate of animal production (e.g., growth, production of milk or eggs) will also affect intake of feed nutrients and subsequent excretion of elements found in manure (e.g., N, C, S, P). In general, higher rates of production result in greater rates of nutrient excretion per animal and lower rates of excretion per unit of animal product. How animals are managed may also affect rates of nutrient excretion. In particular, improved feeding practices increase the efficiency of nutrient utilization and decrease excretion rates. Some feeds are more digestible or metabolizable than others. The proper combination of certain feeds to provide the optimal balance of energy and protein, or complementation of different amino acids, results in more efficient metabolism and lower feeding requirements. Some feed additives improve digestion or metabolism or decrease protein requirements, and thereby decrease emissions of ammonia and methane. Other management factors that affect productivity include the use of hormones to change partitioning of nutrients within the animal, or grouping animals to decrease feed variations among individuals getting the same diet. The efficiency of animal production has increased continuously and is improving still, resulting in lower methane emissions from the animals and lower nutrient transfers to manure (Havenstein et al., 1994).

Unless a process-based modeling approach is used, it is unlikely that any group will be able to develop a model to represent all of the techniques used to improve the efficiency of nutrient use by animals and to incorporate all of the possible current technologies, or to keep up with the development of new technologies. Animal management has a profound impact on the total nutrients transferred to manure and then lost from the farm to pollute air and water. Thus, it is essential that a means be developed to predict these flows to manure. Using the mass balance approach, the quantity of each element (e.g., N, S, C) excreted in manure or volatilized must equal the amount consumed by animals minus the amount transferred productively to meat, milk, or eggs (minus any carbon volatilized directly to the air from the animals) if there is no net accumulation. Since little nitrogen or sulfur is volatilized directly from animals to the air, that which is fed and not accounted for in animal products must leave the animal in manure.

The approaches to models for each pollutant depend on whether the pollutant is important because of long-term global consequences (e.g., methane), short-term local consequences (e.g., hydrogen sulfide), or both (e.g., ammonia), as described previously. In the case of long-term and global predictions, the annual rates of emissions are likely to be adequate, but for local consequences, concentrations (for short-term or chronic exposures) at the farm boundary or nearby homes may be of greater importance.

Predicting Emissions Important on a Global Scale

A system-level approach is recommended for predicting emissions that are important on a global scale (i.e., ammonia, methane, nitrous oxide, nitric oxide). In this case, system level refers to the overall animal production system, which may be contained on several farms (i.e., crops may be produced on a farm different from the one where the animals are housed). The goal is to decrease emissions from the system per kilogram of food (or other product) produced. Individual AFOs may still be regulated in terms of their contribution to system-level emissions, but it would not help to decrease emissions from an AFO only to have them transferred to a different farm, or many smaller farms.

The process-based, mass balance approach begins by predicting nitrogen, carbon, and sulfur in manure by subtracting the quantities of these elements in animal products from the quantities consumed. For every major type of farm animal and every production group within these types, such predictions of intake are already available. Current publications from the National Research Council (NRC, 1994, 1998a, 2000, 2001a) detail nutrient requirements for various animal types and production systems and with varying amounts of production. Moreover, these publications have been updated periodically. Tables D-1 to D–3 in Appendix D are based on the assumption that animals are fed to meet National Research Council recommendations. These tables show that different types of animals convert feed nutrients to human-consumable products at differing efficiencies. The whole-system analysis also requires understanding that cattle, which appear to use feed nutrients least efficiently, in fact consume whole plant feeds (forages) that can be produced with lower environmental impact or by-products that might otherwise be wasted.

To illustrate, nitrogen composition can be determined from the protein content of feed consumed. Thus, for any given AFO, manure nutrient output can be estimated from the number of animals of each type and their average feed consumption and production. Since some farms feed more or less of certain nutrients than the National Research Council recommends, a more accurate estimate of manure output can be made by quantifying the actual feed inputs and the export of animal farm products. Farm feed and export receipts can be used to document this balance, or diets formulated and feed composition can be used. Many producers already maintain good records of the quantities of animal products purchased.

Once the quantity and characteristics of excreted manure have been defined for an AFO, the fate of the elements in that manure has to be predicted. Animal housing, manure collection system, and manure storage facilities affect the transformation of components to volatile forms and the volatilization of manure components. Thus, distribution factors (fractions of manure elements converted to volatile forms) must be determined for different systems, with different types of manure and different climates. Factors would also be needed to predict the fraction of potentially volatile elements of manure that are actually volatilized in different housing or manure systems. These factors might be expressed as functions of exposed surface area in housing and manure storage.

Once the flows of elements to manure storage and the transformations and emissions there have been predicted, the remaining opportunities for emissions from the AFO occur during the removal of manure from storage and its application on crops. These estimates are expressed as a fraction of precursors available for emission from the farm and may be affected by animal type, soil type, cropping system, cropping area, weather and climatic conditions, and concentrations and amounts of the elements in manure.

Ammonia

The committee recommends making estimates and directing control strategies toward decreasing total reactive nitrogen emissions to the environment rather than specifically toward decreasing ammonia volatilization (see section "Reactive Nitrogen Emissions"). However, the prediction of ammonia emissions is necessary for other planning and research purposes. Ammonia emissions on a global scale and extended time frame (e.g., one year) can be predicted from a process-based model that uses mass balances. These long-term emission estimates are adequate for addressing global concerns with ammonia emissions such as nitrogen loading to the environment. However, daily ammonia emissions are also needed to address regional air quality issues such as particle and smog formation and ammonia deposition. Therefore, the committee addresses models for predictions at both annual and daily time frames. Three major steps are needed for a mass balance approach: (1) excretion of manure nitrogen, (2) conversion of manure nitrogen to ammonia, and (3) volatilization of ammonia in housing, from manure storage, and field application.

Total nitrogen excretion from animals, in the absence of changing herd or flock size, is equal to total nitrogen intake (mostly as protein) minus nitrogen in animal products. Most of excreted nitrogen occurs in one of two forms: urea (mammals) or uric acid (birds) in urine and complex organic compounds in feces. When proteins are fed that are not readily digestible, fecal nitrogen increases. When excess amounts of digestible nitrogen are consumed, urine nitrogen amounts increase. Once excreted, enzymes in feces hydrolyze urea and uric acid to ammonia, which can be volatilized from housing (especially at high pH), storage (especially if uncovered), and during land application (especially when not incorporated into soil). Although fecal nitrogen is converted to ammonia more slowly, over long periods of storage (especially in warm climates) considerable fecal nitrogen can be volatilized as NH_3.

Total nitrogen intake can be quantified and documented for an individual AFO by one of three different methods. The easiest approach is to assume that the herd or flock consumes the amount of nitrogen recommended by the National Research Council (1994, 1998a, 2000, 2001a) for its production level. Whereas many producers feed larger or smaller quantities of nitrogen than recommended by the NRC, one of two alternative approaches can be used to quantify nitrogen intake on a particular AFO. Total nitrogen in crops produced on the farm and in the feeds imported can be quantified over a year using farm records. Most producers can estimate quantities of feed produced and keep records of feed purchases. Feed tables can be used to estimate nitrogen content for grains, and chemical analyses are routinely performed on feeds that vary in nitrogen content. An alternative to this approach is to add the nitrogen consumed by each production group on the farm over a specific period of time. In this case, the amount of each diet consumed is multiplied by its nitrogen content. Since diets are carefully formulated for specific crude protein amounts, the nitrogen percentage of each diet is generally available. The quantity of diet consumed is also estimated or measured directly on most farms.

Total net nitrogen in animal products sold can be determined by multiplying the quantity of each product sold by its nitrogen content and subtracting the nitrogen in animals purchased. Total urine and fecal nitrogen can be determined as total nitrogen intake minus net nitrogen in animal products. Fecal nitrogen can be calculated as the indigestible feed nitrogen and endogenous losses as described by current diet formulation models for different species (NRC, 1994, 1998a, 2000, 2001a). Urinary nitrogen is the remainder of the excreted nitrogen. These estimates of nitrogen excretion could be made for longer or shorter time intervals for use in estimating annual or daily ammonia emissions.

Once the total amount of nitrogen excreted has been determined, the rapid ammonia emissions from housing can be calculated as a fraction of the urine nitrogen. For each major type of housing system, coefficients would be needed to predict the percentage of urine nitrogen likely to be converted to ammonia and volatilized for the type of housing and geographic area (for annual emissions) and short-term weather conditions (daily emissions). Limited data are available currently to make such estimates (Muck and Steenhuis, 1982; Monteny and Erisman, 1998).

Nitrogen in manure storage would be estimated from the flows of inorganic and organic nitrogen into storage, measured or assumed removal for field application, and predicted emissions during storage. A typical predictive model for daily or annual ammonia emissions from manure storage would include two modules: ammonia generation in the manure, and ammonia emission from the manure surface. Ammonia in manure is generated partly from urea via enzymatic conversion, which is a rapid process; and partly from the mineralization of organic nitrogen in manure, which is a relatively slow process. Normally, for typical manure collection systems where urine and feces are mixed, urea is assumed to already have been converted to ammonia when the manure reaches storage. Additional ammonia may be generated by mineralization of organic (e.g., fecal, bedding) nitrogen over time and is controlled by the following factors:

- storage time;
- environmental conditions such as temperature; and
- characteristics of the wastes such as biodegradability, moisture content, oxygen content, and pH.

Thus, the amount of ammonical nitrogen in stored manure is a function of the amount and form (urine versus fecal) of nitrogen and other compounds coming into storage; the length of the storage period, and other factors that affect oxygen content, pH, and temperature of the stored manure. The ammonical nitrogen in manure storage at each time could be estimated as the amount at a previous time plus additions from animal housing and mineralization of organic nitrogen in storage, minus removal and nitrogen emissions. The amount of organic nitrogen in manure storage must be estimated because it affects the rate of ammonical nitrogen formation. It would be equal to the previous organic nitrogen plus additions from animal housing, minus losses to inorganic nitrogen and removal for fertilizer application.

Quantitative information on the production of ammonia in animal manures is scarce in the literature. Zhang et al. (1994) developed equations for predicting the production rate of ammonia during storage of swine manure as a function of time and depth in the manure. However, their study did not account for the influence of different temperatures, manure solid content, and oxygen content. More research is needed to develop accurate prediction models for quantifying the production rate of ammonia due to organic nitrogen mineralization in different types of manure management systems.

Once ammonia is generated in animal housing or manure storage, emission of ammonia from the manure to the atmosphere is controlled by the aqueous chemistry of ammonia in the manure and convective mass transfer at the manure surface. The factors that affect emission rates from manure storage include the following:

- water temperature,
- air temperature,
- relative humidity,
- wind velocity, and
- manure characteristics such as pH and solids content.

The rate of emission of ammonia from manure can be calculated as the product of the convective mass-transfer coefficient (K_{LN}) and the concentration of ammonia at the surface layer.

The mass-transfer coefficient K_{LN} is a function of manure temperature, air temperature, wind velocity, and relative humidity. Various equations are available in the literature for K_{LN}. However, most of those equations were developed from controlled experiments with convective mass-transfer chambers and have not been validated well using field-scale experiments. More research is needed to calibrate and validate them. An example is given in Appendix J, which is based on the two-film theory.

The concentration of gaseous NH_3 in air at the manure surface depends on the total concentration of dissolved ammonical nitrogen (NH_3 + ammonium ion [NH_4^+]) at the surface of the liquid manure. Thus, the greater proportions of ammonical nitrogen in the form of NH_3, the greater are the emissions from the solution. The ratio of NH_3 to NH_4^+ depends on the pH of the solution and the equilibrium constant (K_a) of the reaction interconverting NH_3 and NH_4^+. As the pH of the solution increases, more of the ionic form is converted to NH_3 and the emissions of ammonia increase. It has been found by researchers that the K_a in wastewater has a different value from that of pure water (Zhang et al., 1994; Liang et al., 2002). The K_a in animal manure ($K_{a,m}$) is 25-50 percent of the K_a in water, depending on the characteristics of manure, such as solids content. If the stratification of NH_3 in manure is negligible, total ammonical nitrogen can be assumed to be the concentration in the bulk liquid. On any given day, the ammonical nitrogen concentration can be calculated from the concentration at the end of previous day, the change in concentration due to influx from animal housing and the degradation of organic nitrogen in stored manure.

The amount of nitrogen applied to cropland can be predicted as the input to manure storage minus the losses from storage. The amount volatilized depends on the form of nitrogen applied and an additional set of coefficients for different climatic regions, soil types, crops, and management systems (Denmead, 1997).

As with estimates of nitrogen excretion, using on-farm measurements may improve the accuracy of predictions and provide an incentive to decrease emissions by means not considered by regulators. Nitrogen lost from housing and storage can be calculated as nitrogen excreted (as determined previously) minus nitrogen applied to fields from storage (as measured). The nitrogen applied may be determined as manure volume applied times nitrogen concentration (mass per volume) of applied manure. Ammonia emissions from storage can then be estimated as a fraction of nitrogen lost from storage depending on housing and storage type.

Whereas a large amount of the nitrogen excreted by animals is lost from the production system as ammonia, it is vitally important to accurately predict nitrogen excretion from animals. These estimates can be made quite readily for typical farms with typical feeding levels for specific production rates. However, some AFOs use different practices that increase (e.g., overfeeding of protein) or decrease (e.g., feeding specific amino acids) the nitrogen excreted per unit of animal product compared to typical farms. Therefore, calculation of manure production rates from feed nitrogen and product nitrogen may improve the estimate of nitrogen emissions. On the other hand, although sampling and analysis can be used to estimate the manure nitrogen applied, many manure systems are difficult to sample and such measurements may not be useful in some cases. Most of the nitrogen lost from animal production systems is volatilized ammonia, and it can be quantified as nitrogen emissions. Therefore, mass balance-based prediction models can be used to identify farms according to their management practices and to quantify the impact of management techniques on ammonia emissions. These predictions could be improved by determining the percentage of nitrogen emissions as ammonia, as opposed to molecular nitrogen (N_2) or other forms of nitrogen, for different management systems.

Prediction of ammonia emissions in animal housing, manure storage, and field application is feasible in the near term (i.e., five years) using a process-based model. However, as discussed in the section "Reactive Nitrogen Emissions," this may not be the best option for predicting emissions on individual farms for regulatory or incentive programs.

Molecular Nitrogen, Nitrous Oxide, and Nitric Oxide

As with ammonia, the committee recommends predicting and directing control strategies toward decreasing total reactive nitrogen emissions to the environment rather than specifically toward nitrous oxide and nitric oxide (see discussion of reactive nitrogen emissions). However, prediction of global nitrous oxide (N_2O) and nitric oxide (NO) emissions is needed for other planning and research purposes. As with NH_3, emissions of other nitrogen-containing compounds can best be estimated as a fraction of nitrogen excreted to emittant (Müller et al., 1997). Many other factors, including soil moisture, compaction, and pH, also contribute substantially to these emission rates. As a general rule, N_2O is 3 to 5 percent of available nitrogen emitted from manure or other nitrogen fertilizer applied to soil (Hansen et al., 1993; Flessa et al., 1996; Flessa and Beese, 2000). These emissions derive from fields used to produce feed for animals using inorganic forms of nitrogen fertilizer as well as animal manures. In addition, 1 to 2.5 percent of the nitrogen that leaves the farm as volatilized ammonia or nitrogen oxides, or leaches or runs off as nitrate, volatilizes as nitrous oxide elsewhere in the environment (IPCC, 2000).

Although N_2 gas is not harmful, estimation of its emissions from manure handling and cropping systems would improve our ability to predict total nitrogen emissions and to develop better management strategies. Nitrogen gas is formed from nitrate in an anaerobic process called denitrification after nitrate has been produced from ammonia in an aerobic process called nitrification (Thompson et al., 1987). Thus, N_2 may evolve from manure storage structures or soils where both anaerobic and aerobic compartments exist. Research is needed to determine the ratio of N_2O to N_2 and the factors that affect it because generation rates of both gases are linked to rates of nitrification and denitrification (Abbasi et al., 1997), and management may be able to shift reactions to favor the benign product N_2 (Dendooven et al., 1996).

Reactive Nitrogen Emissions

The committee has recommended in several places in this report that EPA and USDA take a systems approach to develop control strategies aimed at decreasing undesirable emissions from AFOs. With regard to ammonia, N_2O, and NO, a systems approach would help optimize the animal production systems to decrease the overall environmental impact of all of these emissions. An example of the opposite would be to promote strategies to decrease ammonia emissions that simultaneously increased N_2O emissions. A systems approach is also needed to decrease total reactive nitrogen emissions to air and water simultaneously. Water quality regulations for AFOs might require nitrogen-based nutrient management plans. In many situations, the most cost-effective plans would be those that result in the greatest volatilization of ammonia to the atmosphere in order to decrease the need to "dispose" of excess nitrogen from the AFO in the form of water-soluble nitrate. A systems approach is also needed to decrease total emissions of reactive nitrogen per unit of animal product, rather than simply decreasing reactive nitrogen emissions from individual AFOs. Control strategies to decrease nitrogen emissions from AFOs might inadvertently provide incentives for producers to ship their problems to other farms, rather than improving the efficiency of the whole animal production system. For example, nitrogen emissions from an AFO can be decreased by purchasing more feed and by shipping manure off-farm. In this way, nitrogen emissions will occur from the farms that grow the crops or apply the manure, rather than from the AFO, but the emissions of nitrogen to the environment may not be affected. Because of the importance of the systems approach, the committee recommends using a farm nitrogen balance as described below.

FINDING 10. A systems approach, which integrates animal and crop production systems both on and off (imported feeds and exported manure) the AFO, is necessary to evaluate air emissions from the total animal production system.

RECOMMENDATION: Regulatory and management programs to decrease air emissions should be integrated with other environmental (e.g., water quality) and economic considerations to optimize public benefits.

Reactive nitrogen in the form of NH_3, N_2O, or NO may be lost to the atmosphere; as soluble nitrogen running off into surface water, or as nitrate leaching into groundwater. Some nitrogen may be converted back to harmless N_2 gas. The total emission of nitrogen from the farm (N_{loss}) is therefore composed of those environmentally destructive emissions to air (N_{air}) and water (N_{water}) and the harmless conversions to N_2:

$$N_{loss} = N_{air} + N_{water} + N_2. \quad \text{(Eq. 5-2)}$$

Measurements of individual emissions (e.g., ammonia volatilization, N runoff, and nitrate leaching) are difficult and expensive, leading to the predicament that few data are available on which to base mathematical models for predicting individual emissions. In contrast, the aggregate of all nitrogen emissions to the environment can be predicted from reliable measurements that can be documented on individual farms based on nitrogen inputs (N_{input}) in imported feed, legumes produced on-farm, and imported fertilizers and on nitrogen exports (N_{output}) in animal products and exported feeds:

$$N_{loss} = N_{input} - N_{output}. \quad \text{(Eq. 5-3)}$$

Not all of these nitrogen emissions from the production system are destructive to the environment because some N_2 gas is also formed. The reactive nitrogen emissions can be calculated by subtracting estimates of N_2 gas formed for different types of management from the total nitrogen emissions. For most animal production systems, only a small percentage of the nitrogen that is not accounted for is thought to be lost as N_2. In soils, N_2 emissions are associated with N_2O emissions (Abbasi et al., 1997). Emissions of N_2 can occur from anaerobic or aerobic lagoons (most are a combination of both), biogas generation and combustion, constructed or natural wetlands, and cropping systems. Little is known about how much of the nitrogen lost from these systems is converted to N_2 gas and how the systems can be optimized to increase N_2 as they decrease other nitrogen emissions. Nonetheless, in the near term, predicting total reactive nitrogen emissions from individual AFOs, using a mass balance approach, is likely to be less costly and more accurate than attempting to estimate the individual types of nitrogen emissions from measurements of concentrations and flow rates (Box 5-1).

Control strategies for nitrogen emissions should consider using estimates of total reactive nitrogen emissions. These strategies can include both performance standards based on predicted reactive nitrogen emissions for production systems and technology standards that can be documented to be effective using the models of reactive nitrogen emissions. For individual AFOs, reactive nitrogen emissions can be predicted and documented (Dou et al., 1996). Total nitrogen inputs can be quantified and documented using nitrogen compositions, feed and bedding purchases, fertilizer imports, estimated legume nitrogen fixation (legume production divided by an efficiency factor), and imported animals. (Another input includes atmospheric nitrogen deposition, but this cannot be controlled by AFO operators.) Known farm nitrogen outputs can be quantified as feeds sold, and animals and animal products shipped, each times its nitrogen content. The emissions of N_2 currently cannot be predicted accurately because of a lack of research in this area, but with new research on different types of manure and cropping systems such predictions may become feasible in the near term.

Control strategies to decrease reactive nitrogen emissions to the environment are likely to be applied to individual AFOs. However, the objective is to decrease emissions from the entire animal production system, which includes off-farm crop production and manure handling. If an AFO exports manure or imports crops, its predicted reactive nitrogen emissions will decrease, and those of the farms handling the manure or producing the crops will increase. The goal of control strategies should be to provide incentives for improving the efficiency of nitrogen utilization among all contributors to the production system. Incentives can be created to encourage AFOs to decrease these predicted reactive nitrogen emissions to the environment as a fraction of animal products produced. For an individual AFO, the reactive nitrogen emission to the environment can be predicted. In order to calculate the reactive nitrogen losses from the entire production system, after calculating the reactive nitrogen emission from the AFO, typical emissions for production of imported feeds or exported manure can be added. This process would put all AFOs on a similar footing.

FINDING 11. Nitrogen emissions from AFOs and total animal production systems are substantial and can be quantified and documented on an annual basis. Measurements and estimates of individual nitrogen species components (i.e., NH_3, N_2, N_2O, NO) should be made in the context of total nitrogen losses.

BOX 5-1 Sample Calculations of Whole-Farm Nitrogen Balance

Nitrogen may leave an AFO when ammonia volatilizes, inorganic or organic nitrogen runs off from fields or holding areas, nitrate leaches into groundwater, or nitrous oxide volatilizes. In addition, some molecular nitrogen may be released from manure storage or treatment facilities, wetlands, or fields. Although it is very difficult to estimate any of these potential emissions of nitrogen from individual AFOs, the aggregate of all emissions can be estimated reliably as the nitrogen balance. To identify weak points on the farm, the efficiency of farm components must be determined and compared to alternative practices.

Dou et al. (1996) describe software to estimate nitrogen balances on farms retrospectively and to predict future nitrogen balances based on different management decisions. Using this software, the nitrogen balance results for recommended management practices can be compared to results from observed practices.

Sample nitrogen balances were calculated for two Pennsylvania dairy farms (Dou et al., 1998). One farm had 109 lactating Holstein cows, 29 dry cows and bred heifers, and 84 unbred replacement heifers or calves. Manure was scraped every other day and stored in a concrete pit for up to a year. Soils ranged from low to medium productivity, and the crop area and total crop dry matter (DM) yield per year were 69 ha (hectares) of corn silage (320 Mg DM/yr), 33 ha winter rye haylage (93 Mg DM/yr), and 23 ha alfalfa hay or haylage (86 Mg DM/yr). The farm purchased hay (89 Mg DM/yr) soybean meal (55 Mg DM/yr), corn grain (102 Mg DM/yr), corn distiller's grain (34 Mg DM/yr), and whole cottonseed (42 Mg DM/yr).

Nitrogen contents of milk, live animals, crops, and feeds were determined by using laboratory analyses that were already employed for diet formulation and milk pricing or by using book values provided by the software (Dou et al., 1996). The quantities of all imports and exports were obtained from receipts for purchased feeds or fertilizers and milk and animal sales. Legume nitrogen fixation was estimated as a fraction (0.6) of the legume nitrogen harvested on the farm, with the remainder (0.4) coming from crop or manure residues. The quantity of nitrogen in imports and exports is shown in

RECOMMENDATION: Control strategies aimed at decreasing emissions of reactive nitrogen compounds (Nr) from total animal production systems should be designed and implemented now. These strategies can include both performance standards based on individual farm calculations of nitrogen balance and technology standards to decrease total system emissions of reactive nitrogen compounds by quantifiable amounts.

the table below. The difference between imports and exports (N balance) was 12.5 Mg N/yr which represents the emissions of nitrogen to the environment or accumulated in soil.

Source of N Input	Quantity (Mg N/yr)	Source of N Output	Quantity (Mg N/yr)
Purchased feed	11.8	Milk sold	3.9
N fertilizer	3.3	Animals sold	0.6
Imported bedding	0.3	Crops sold	0.1
Biological N fixation	2.4	Manure exported	0.4
Total N imported	**17.8**	**Total N exported**	**5.0**
	N Balance	12.8 (MgN/yr)	

It may be useful to calculate the nitrogen utilization efficiencies of the different farm components (e.g., herd, manure storage and application, crops). The efficiency of herd production is calculated as nitrogen in milk and live animals produced (3.9 + 0.6) divided by nitrogen intake (11.4, data not shown). In this case, it was 0.22. This estimate was compared to an alternative diet formulation that would have provided an efficiency of 0.27. The improved diet formulation was expected to increase imported feed nitrogen as well as milk production, so the total nitrogen emissions to the environment would not be changed. However, with the higher milk production, total emissions from the farm per unit of milk produced were lower, and the environmental impact would be favorable.

Manure storage efficiency is calculated as nitrogen applied to crops divided by nitrogen excreted, where nitrogen excreted equals nitrogen intake minus that exported in milk, live animals, and crops. In this case, nitrogen excreted plus bedding nitrogen was 17.5 Mg/yr and the amount measured as applied on crops was 8.5 Mg/yr for a storage efficiency of 0.66. Using the Dou software (Dou et al., 1996), the quantity of manure nitrogen application was estimated as 10.0 Mg/yr, or 0.63 times manure nitrogen excreted. In this case, emissions of nitrogen from manure storage could be estimated for different types of storage systems, and the impact on the whole farm could be calculated.

Enteric Methane Production

For all of the compounds considered in this report, only methane is known to be emitted directly from animals, and methane is produced in significant quantities only in cattle or other ruminant animals. The EPA (2001a) emission factor approach did not consider enteric methane emissions, although these account for the majority of methane emissions from agriculture. In the United States, enteric emissions account for 19 percent and anaerobic manure lagoons account for 13 percent of all anthropogenic methane emissions (EPA, 2000a). Globally, about six times more methane is estimated to come from enteric emissions than from decomposition of animal manure, but if anaerobic manure storage became more common, the latter source would increase (Johnson and Ward, 1996).

Methane emissions from ruminant livestock in the United States currently are estimated by EPA by (1) dividing animals into homogeneous groups; (2) developing emission factors for each group; (3) collecting population data; (4) multiplying the population of each group by its emission factor; and (5) summing emissions across animal groups and geographic regions (EPA, 1993a). Among livestock, cattle are examined more closely than other animals because they are responsible for the majority of U.S. livestock emissions.

Johnson and Johnson (1995) used a more accurate and robust approach based on energy consumption by cattle. Methane production was estimated to be 6.0 percent of dietary gross energy consumption for cow-calf, stocker, or dairy systems and 3.5 percent for feedlots. This approach was based on research demonstrating the similarity for different breeds and stages of cattle production but accounts for methane inhibition with the high-concentrate diets fed to feedlot cattle.

Other approaches considering feed characteristics may improve the accuracy of these estimates but also increase the complexity. Blaxter and Clapperton (1965) developed an equation based on a series of methane production measurements from sheep fed different diets. Methane was predicted from the digestible energy (rather than the gross energy) and the energy intake relative to maintenance requirements. Moe and Tyrrell (1979) derived an equation that predicted enteric methane from diet composition, based on measurements made on cattle fed high-quality dairy diets:

$$\text{Methane (megajoules per day)} = 3.406 + 0.510\,(\text{cell solubles}) + 1.736\,(\text{hemicellulose}) + 2.648\,(\text{cellulose}). \qquad \text{(Eq. 5-4)}$$

The number of megajoules refers to the amount of heat that could be released if all of the enteric methane produced were burned to carbon dioxide and water. Wilkerson et al. (1994) concluded that this methane prediction equation resulted in the lowest prediction error of six equations that were compared.

Methane Emissions from Manure Storage

Methane is produced from microbial decomposition of animal manure under anaerobic conditions. The major factors that determine methane production are

- types and populations of microorganisms present;
- storage or retention time;

- environmental conditions such as temperature; and
- characteristics of the wastes such as biodegradability, nutrient availability, moisture content, oxygen content, and pH.

Many microorganisms are involved in the decomposition of animal manure, including fungi, bacteria, and protozoa. Anaerobic decomposition that leads to methane production involves hydrolytic, acid-forming, and methanogenic organisms. The presence and population of methanogenic bacteria are important for methane production. Manure from ruminants contains more methanogenic bacteria than manure from nonruminant animals. Temperature affects the rate of biochemical reactions, the types of functioning microorganisms, and therefore, the rate of methane production. Methanogenesis in animal manure has been observed in a large temperature range of 4-75°C. Methane production generally increases with increasing temperature. The characteristics of animal manure are governed by animal diet and by manure handling, collection, and storage methods. The greater the energy content and biodegradability of the feed, the greater is the methane production potential of the manure. For example, manure from animals fed with grain-based, high-energy diets is more degradable and has higher methane production potential than manure from animals fed with a roughage diet. All manure has a maximum (or ultimate) methane production potential, which is determined by its chemical composition. The maximum methane production potential is defined as the quantity of methane that can be produced per unit mass of volatile solids (VS) in the manure and is commonly expressed as B_0 with a units of cubic meters of CH_4 per kilogram of VS. Volatile solids, also called organic matter, are approximately 55 percent carbon on a mass basis for manure.

In theory, the maximum (ultimate) methane production capacity of a quantity of manure can be predicted from the gross elemental composition. In practice, however, insufficient information exists to implement this approach, and direct laboratory measurement is recommended for determining methane production capacity (EPA, 1992). The effects of animal diet on the maximum methane production potential are illustrated in Table 5-1.

Microbial growth requires nutrients such as nitrogen, phosphorus, and sulfur. Animal manure typically contains sufficient nutrients to support microbial growth so that nutrient availability is not a limiting factor in methane production under most circumstances. Moisture content relates to the availability of water to the microorganisms and the ability to maintain an oxygen-free environment in the manure. More than 80 percent moisture content is conducive to methane production. Animal manure as excreted contains 70-91 percent moisture or 9-30 percent total solids (MWPS, 1985). Its moisture content may change to higher or lower amounts depending on the methods of manure collection, handling, and storage in the animal waste management systems. Based on the total solids (TS) content of manure and the requirements for different handling methods, animal waste management systems can be classified into three types: (1) solid system (TS > 20

TABLE 5-1 Maximum Methane Production Potential of Animal Manure as Affected by Different Diets

Animal Type	Diet	Methane Yield B_0 (m^3/kg VS)[a]
Swine	Corn-based, high-energy diet	0.44-0.52
	Barley-based diet	0.36
Dairy cattle	58-68% silage	0.24
	72% roughage	0.17
Beef cattle	Corn-based, high-energy diet, manure collected from concrete	0.33
	7% corn silage, 87% corn, manure collected from dirt lot	0.29
	91.5% corn silage, 0% corn, manure collected from dirt lot	0.17
Poultry, caged layer	Grain-based diet	0.39

[a]Volume of methane under standard conditions of 1 atmosphere of pressure and 25° C.

SOURCE: Safley and Westerman (1990).

percent); (2) slurry system (TS = 10-20 percent); and (3) liquid system (TS < 10 percent) (MWPS, 1985). Examples of different systems are solid manure storage, dry lot, deep-pit stacking, and litter for solid manure management; under-floor deep-pit storage and slurry storage for slurry systems; and anaerobic lagoons for liquid systems. Methane production potential increases in sequence from solid manure to slurry manure to liquid manure systems. Methanogenic bacteria are obligate anaerobes (i.e., they require an absolutely oxygen-free environment to survive). They are also sensitive to low pH of the manure. The optimum pH is near 7.0, but methane can be produced in a pH range from 6.2 to 8.5, and a pH outside the range 6.8-7.6 will decrease methane production. The retention time of manure in waste storage systems is also an important factor. If all the other conditions are the same, longer times will lead to more methane production.

The maximum methane production potential (B_0) can be determined in the laboratory using anaerobic digestion assay. For the purposes of municipal wastewater treatment, an assay called Biochemical Methane Potential is commonly used by researchers and wastewater plant operators (Speece, 1996). The assay uses 100-mL serum bottles, which are incubated at 35°C, as the testing reactors. Because of the presence of relatively large particles and the nonhomogeneous nature of animal manure, larger reactors are recommended to ensure the accuracy of measurements. A standard anaerobic digestion assay for determining B_0 for animal manure still has to be developed. Laboratory protocols used by animal manure researchers typically use batch digestion of 5 to 15 g VS of manure in an anaerobic reactor of 1- to 3-L volume at 35°C and the measurement of hourly or daily methane production. When the manure stops producing methane, which means that it has been exhausted of biodegradable organic matter, the total volume in liters of methane produced during the whole testing period is divided by the initial VS in grams in the manure to give the maximum methane production potential (B_0, liters of CH_4 per gram of VS or cubic meters of CH_4 per kilogram of VS). The anaerobic reactor must contain sufficient bacterial culture that has been well adapted to the manure to be tested. The actual methane production in manure storage is lower than the maximum methane production potential.

Actual methane production from manure may be estimated from the maximum methane production potential with considerations of animal waste characteristics, manure storage type and time, and climatic conditions. The following methane emission estimation procedures are adapted from EPA (1992), which is used to estimate emissions from livestock and poultry manure. A similar approach has also been adopted by the Intergovernmental Panel on Climate Change (IPCC) for estimating methane emissions from animal manure (IPCC, 2000). The methane emission production on a livestock farm may be calculated using the following equation:

$$TM_{CH_4} = \Sigma VS_i \bullet B_{0i} \bullet MCF_j \bullet WS_{ij} \bullet CAF, \qquad \text{(Eq. 5-5)}$$

where

TM_{CH_4} = total annual methane emission from manure storage on the farm in cubic meters of CH_4 per year;
VS_i = volatile solids produced annually by animal i in kilograms;
B_{0i} = maximum methane production potential of the manure from animal i in cubic meters of CH_4 per kilogram of VS;
MCF_j = methane conversion factor for manure storage j, which represents the extent to which B_0 is realized (*note:* $0 \leq MCF \leq 1$);
WS_{ij} = fraction of animal i's waste handled in the manure storage j; and
CAF = climate adjustment factor for the farm, which represents the extent to which B_0 is realized under climatic conditions (e.g., temperature, rainfall) on the farm (*note:* $0 \leq CAF \leq 1$).

The MCF varies for different types of manure storage and may assume a value from 0.1 for solid manure storage to 0.9 for anaerobic lagoons (Chen et al., 1988; EPA, 1992). The moisture content and retention time of manure in storage affect the MCF. A relationship between the MCF and manure storage conditions, such as moisture content or total solids concentration and retention time may be developed. Research is needed to determine MCF for different types of manure management systems. The CAF mainly varies with ambient temperature. The effect of temperature on methane production from animal manure has been extensively studied (Hobson, 1990), and mathematical relationships have been developed for predicting the effect of temperature on methane production in certain types of anaerobic processes, including anaerobic lagoons. Most research has focused on mesophilic (25-40°C) and thermophilic (40-65°C) temperature ranges (Hobson, 1990) for the purpose of producing methane from animal manure. The temperature of manure storage systems is mostly in the psychrophilic temperature range (less than 25°C). Safley and Westerman (1988) and Safely (1992) published the relationships between the temperature and methane yield of dairy and swine manure using laboratory digesters operated at different degrees ranging from 10 to 23°C. The information was used to predict methane production from anaerobic lagoons. More research is needed to correlate the CAF to the ambient temperatures for different types of manure management systems.

Predicting Emissions Important on a Local Scale

Odor

The odor of animal manure is complex both because of the large number of compounds that contribute to it and because it involves a subjective human response. Much research has been conducted to identify odorous compounds in animal manure. More than 75 compounds were initially identified (Miner, 1975; Barth et al., 1984). Recently, Schiffman et al. (2001) identified 331 odor-causing compounds in swine manure. Odorous compounds include ammonia, amines, sulfides, mercaptans, organic acids, alcohols, aldehydes, esters, and carbonyls. The volatile fatty acids (VFAs), phenols (phenol and *p*-cresol), and indoles have been considered to be the major odorous compounds (Schaefer, 1977; Spoelstra, 1980; Williams, 1984).

The odor intensity and characteristics of animal manure change over time and are affected by manure handling and storage conditions. For example, in fully aerated manure, biochemical oxidation of various organic compounds occurs. Chen et al. (1994) reported that aeration of swine manure degrades VFAs, phenol, *p*-cresol, and skatol within 24 hours. Aerobic treatment is known for its effectiveness in decreasing manure odors. In partially aerated manure where low dissolved oxygen limits aerobic organisms, the production of odorous compounds is highly related to redox potential (ORP) in the manure. Beard and Guenzi (1983) studied the evolution of sulfur gases, including hydrogen sulfide, dimethyl sulfide (DMS), methanethiol (MeSH), carbonyl sulfide (COS), and carbon disulfide (CS_2), in cattle manure at different ORP levels ranging from –200 to +300 mV (millivolts) and found that the gases produced varied with the ORP. For example, DMS and dimethyl disulfide (DMDS) were highest at the ORP of 0 mV, while hydrogen sulfide (H_2S) and MeSH were greatest at an ORP of –100 mV or less. Under anaerobic conditions, if the environment is conducive to acetogens and methanogens, the degradation of manure is more complete, and gases such as methane, carbon dioxide, ammonia, and hydrogen sulfide are the main gaseous products. However, if the environment is not suitable for bacteria to carry out complete degradation, other bacterial metabolic processes can produce many odorous compounds. Generally speaking, more completely degraded manure, (e.g., in anaerobic digesters or lagoons), will give off fewer odors than less completely degraded manure (e.g., in manure storage pits or tanks) (Pain et al., 1990). The major factors that can affect the anaerobic degradation process include pH, temperature, solids content, and presence of inhibitory substances such as high concentrations of ammonia and VFAs. Odor-causing compounds must be water soluble to reach the olfactory nerve; so using impermeable covers on manure storage will retard the release of these compounds and aid in controlling odors.

Understanding the production rates of odorous gases in manure with different characteristics and storage conditions is critical for predicting odor emission rates. More research is needed in this area. Correlating types and concentrations of odorous gases and their mixtures to the odor strength measured by human olfactory responses (e.g., using the olfactometry method) is another critical step. Some effort has been made to correlate individual gases to odor concentration. Schaefer (1977) correlated 13 compounds with odor intensity measured by a mobile olfactometer and found that odor intensity had the highest correlation with *p*-cresol. Spoelstra (1980) and Williams (1984) reported that VFAs could be used as a significant indicator of manure odor. More recently, Hobbs et al. (2001) developed a correlation between human olfactory response and an odor mixture of four gases, including hydrogen sulfide, 4-methylphenol, ammonia, and acetic acid, and indicated that determining the odor intensity of pig manure using these main odorants would be a suitable approach.

Until a better understanding of odor generation related to the manure characteristics and environmental conditions in various types of manure storage is achieved and quantitative relationships between various odorous gases as well as their mixtures, and odor intensity have been established, prediction of odor emission rates from manure storage is not possible. At present, conducting on-site measurement of odors is the only available approach for quantifying odor emission rate from a particular farm or manure facility.

Particulate Matter

As discussed in Chapter 3, primary particulate matter emissions in animal feeding operations result from mechanical generation and entrainment. In cattle feedlots, the activity level of the cattle, along with the moisture content of the ground, affect the amount of particulate emissions. Particulate emissions from enclosed animal houses also depend on building ventilation rates and whether the ventilation is natural or mechanical. On-farm sources of particulate emissions may also include unpaved roads, grain mills or storage facilities, crop production equipment, and feeding equipment.

Unlike nitrogen emissions, particulate emissions do not lend themselves to a process modeling approach based on conservation principles. Moreover, there is little information published on particulate emission rates from various components of AFOs. For example, particulate emissions from cattle feed yards are affected by the animals' level of activity, which varies during the day, and may be a function of other parameters such as ambient temperature or density of animals per paddock. For other farm components, models are already available for predicting particulate emissions. For example, models to predict emissions given the silt content of the road surface, mean vehicle speed, mean vehicle weight, mean number of wheels per vehicle, and precipitation are available for particulate emissions from unpaved roads (e.g., Fitzpatrick, 1987). Regression analysis might be used to develop models to similarly calculate emission factors from various AFO components once the key variables have been identified. Given the paucity of research on particulate emissions from AFOs, this effort will require new data on emission rates from AFO components.

Hydrogen Sulfide

Hydrogen sulfide (H_2S) is produced by sulfur-reducing bacteria during anaerobic decomposition of animal manure. The major factors that influence its production rate are

- population of sulfur reducing bacteria;
- amount of sulfur in the manure;
- characteristics of the manure such as moisture content, oxygen content, and pH;
- environmental conditions such as temperature; and
- manure storage time.

The sulfur content of animal manure is directly related to the animals' diet. The major sources of sulfur in animal diets include sulfur-containing amino acids (methionine, cystine, and cysteine) and water-soluble vitamins (biotin and thiamine), as well as some inorganic sulfate. Sulfur is excreted as sulfate, sulfide, and organic compounds in both feces and urine. Reduction of inorganic sulfate to sulfide occurs to a limited extent in nonruminants (Kline et al., 1971) but is prevalent in the anaerobic conditions of the rumen and hind gut of other herbivores. Therefore, for swine and poultry it appears that most of the production of hydrogen sulfide and other volatile sulfur-containing gases occurs as a result of manure decomposition during storage. Emissions may directly occur from a ruminant, but no data are available to date. The characteristics of and environmental conditions in the manure affect the production rate of H_2S. Beard and Guenzi (1983) found that in cattle manure slurry, the level of redox potential affects the production of hydrogen sulfide and other sulfur gases. Their experimental results indicated that production of H_2S was high at the ORP level below –100 mV. Production of hydrogen sulfide at other ORP levels has also been also reported (Farwell et al., 1979).

Hydrogen sulfide likely originates mainly from microbial reduction of sulfate, although it can also be produced by microbial degradation of cysteine and cystine (Freney, 1967; Riviere et al., 1974). Sulfate-reducing bacteria are present in animal manure, but their populations in different manures are not well understood. Riviere et al. (1974) found that the sulfate-reducing bacteria were present in swine manure in amounts up to 104/mL.

Sulfate in the water supply is another source of sulfur in manure. It is reported that the sulfate concentration in the U.S. water supply varies greatly from 0 to 770 mg/L (AWWA, 1990). The relationship between high sulfate concentration in manure and the production of hydrogen sulfide is well demonstrated by Arogo et al. (2000), who studied initial sulfate concentrations (5.89-275 mg/L) in swine manure during a 30-day storage period. Based on biochemistry, sulfide production from sulfate is faster than from organic sulfur. In waste management systems where a large quantity of water is added to animal manure, sulfate in the water supply can be an important source of sulfur.

In aqueous solution such as liquid manure, hydrogen sulfide maintains equilibrium with bisulfide (HS^-) and sulfide (S^{2-}) ions. The presence of hydrogen sulfide as a fraction of total sulfide is affected mainly by pH and temperature. Studies have found that ionization in animal manure differs from ionization in water. Solids content of animal wastewater may also be a factor. The pH of animal waste has a large influence on the hydrogen sulfide fraction. Below pH 5, the H_2S fraction in total sulfide is 100 percent; at pH 7, the fraction is about 50 percent; and above pH 10, the fraction is 0.1 percent.

Emissions of hydrogen sulfide rise with the concentration. For a given amount of total sulfide, lower pH will result in a higher emission rate. Any effort to lower the pH of animal manure that contains hydrogen sulfide and/or has conditions conducive to its production will raise emissions of H_2S. Emission of hydrogen sulfide, like other soluble gases such as ammonia, involves diffusion in wastewater to the water surface and release into the atmosphere. The major factors controlling the emission rate of hydrogen sulfide include air temperature, wind speed, water temperature, and surface area of wastewater. For the same volume of wastewater, a larger surface area will result in higher emissions. Therefore, a waste storage structure with a larger depth and smaller surface area will have a lower emission rate than a structure with a smaller depth and larger surface area. This also explains why agitation of manure in storage tends to increase emissions of hydrogen sulfide. Avery et al. (1975) studied H_2S production in swine confinement finishing buildings and found that it was highly correlated with air temperature, ratio of pit surface area to pit volume, air exchange rate of the building, and daily dietary sulfur intake. Banwart and Bremner (1975) studied the production of H_2S and other sulfur gases from swine manure and found that the volatile sulfur gases produced were predominantly hydrogen sulfide and methyl mercaptan (80 percent) under simulated anaerobic storage conditions. However, during a one-month incubator period, only 0.03 percent of the total sulfur present in swine manure was volatilized. Beard and Guenzi (1983) studied the production of hydrogen sulfide and other sulfur gases in cattle manure slurry and found that sulfur gases volatilized during a three-week period removed 1.7 percent of the total sulfur in the manure. This points out the difficulties in using mass balance approaches for predicting emission of hydrogen sulfide.

Production and emission of gases, including hydrogen sulfide, from manure storage are dynamic processes. Prediction of the emission rate of hydrogen sulfide involves prediction of the production rate of H_2S in the manure and of mass transfer of H_2S from the manure into the atmosphere. Based on the aqueous chemistry of hydrogen sulfide, the mass-transfer processes and governing parameters for H_2S are the same as for ammonia. However, Ni et al. (2000, s2000d) observed that H_2S could be released from manure in bursts, which were characterized by a sudden increases of hydrogen sulfide concentrations in the swine building monitored. It appears that hydrogen sulfide emission shares the characteristics of both ammonia (continuous) and methane (bubbles or bursts) emission. The emission of H_2S from manure storage is a complicated and poorly understood phenomenon.

Production of Hydrogen Sulfide in Animal Manure

Because of the potential health effects of hydrogen sulfide, prediction models for H_2S should have a time scale of hours or days. An approach similar to that for methane production can be used to predict hydrogen sulfide production, except that total sulfur content is used as the manure characteristic instead of volatile solids, a time factor is added to account for the effect of storage time, and the time scale should be on daily basis. The production rate of hydrogen sulfide may be predicted using the following equation:

$$TM_{S^-,k} = \Sigma S_i \bullet S_{0i} \bullet SCF_j \bullet WS_{ij} \bullet CAF, \qquad \text{(Eq. 5-6)}$$

where:

$TM_{S^-,k}$ = total sulfide production from manure storage on the farm in grams of sulfide, for day k;

S_i = sulfur excreted by animal i, plus sulfur per animal added from the water supply, in grams per day;

S_{0i} = maximum sulfide production potential of the manure from animal i in grams of sulfide per gram of sulfur;

SCF_j = sulfur conversion factor for manure storage j, which represents the extent to which S_0 is realized (*note:* $0 \le SCF \le 1$);

WS_{ij} = fraction of animal i's waste handled in the manure storage j;

CAF = climate adjustment factor for the farm, which represents the extent to which B_0 is realized under climatic conditions (e.g., temperature, rainfall) on the farm (*note:* $0 \leq \text{CAF} \leq 1$); and

Quantitative information in the literature on production of hydrogen sulfide in animal manures is limited. Arogo et al. (2000) developed equations for predicting the production rate of hydrogen sulfide in swine manure storage as a function of time and depth in the manure. However, their study did not account for the influence of different temperatures, manure solid content, and oxygen content. More research is needed to develop accurate prediction models for quantifying the production rate of hydrogen sulfide from different types of manure management systems.

Emission of Hydrogen Sulfide from Animal Manure

Continuous emission of hydrogen sulfide from manure to the atmosphere is controlled by the aqueous chemistry of hydrogen sulfide in the manure and convective mass-transfer mechanisms at the manure surface. The pH, manure temperature, air temperature, wind velocity, and relative humidity are major factors that affect the emission process. The pH controls the partitioning of sulfide among three species, H_2S, HS^-, and S^{2-}. The emission rate ($M_{H_2S,k}$) of hydrogen sulfide from manure on the kth day can be calculated using the following mass-transfer equation:

$$M_{H_2S,k} = K_{LS}\ \alpha\ [S^-]_{total,k} \qquad \text{(Eq. 5-7)}$$

where K_{LS} is the convective mass transfer coefficient and $[S^-]_{total}$ is the total sulfide concentration at the manure surface. If the stratification of total sulfide is negligible, $[S^-]_{total}$ can be assumed to be the total sulfide concentration in the bulk liquid. On any given day, the $[S^-]_{total,k}$ can be calculated by the concentration at the end of the previous day $[S^-]_{total,k-1}$, and the new concentration generated, which can be calculated from generation rate $TM_{S^-,k}$ divided by the volume of the liquid in storage (V), as shown below,

$$[S^-]_{total,k} = [S^-]_{total,k-1} + TM_{S^-,k}/V. \qquad \text{(Eq. 5-8)}$$

In Equation 5-7, α is a fraction and can be calculated from the pH and ionization constants $K_{s,1}$ and $K_{s,2}$ (Arogo et al., 1999):

$$\alpha = (10^{-pH})^2/[(10^{-pH})^2 + K_{s,1}\,(10^{-pH}) + K_{s,1}K_{s,2}], \qquad \text{(Eq. 5-9)}$$

where $K_{s,1}$ is the ionization constant for the equilibrium reaction $H_2S = H^+ + HS^-$, and $K_{s,2}$ is the constant for $HS^- = H^+ + S^{2-}$. Their relationships to the temperature in aqueous solutions are well defined. However, the influence of manure characteristics, such as solids content and the presence of metals, is not well understood. In contrast to ammonia, little research has been conducted to quantify the differences between the ionization constants of sulfide in animal manure versus water. The mass-transfer coefficient K_{LS} is a function of manure temperature, air temperature, wind velocity, and relative humidity. Arogo et al. (1999) developed a correlation for K_{LS} for under-floor pit manure storage. More research is needed to define K_{LS} for outside manure storage.

SUMMARY

Whereas it is difficult and expensive to measure most emissions from AFOs directly, and the measurements may be of questionable accuracy, regulatory and management agencies seeking to mitigate emissions need reasonably accurate methods to estimate and attribute them to particular enterprises or activities. One approach is to develop emission factors for different animal production sectors, and estimate emissions as the product of the specific factor for that sector and the number of animals associated with the enterprise or geographic region. The committee found and reported in its Interim Report (NRC, 2002a) that existing data were not adequate at this time to determine accurate emission factors, and that a greater number of emission factors than originally proposed would be needed to explain the variation in emissions from different enterprises. The committee debated the cost and time requirements to conduct necessary studies and develop appropriate emission factors. The approach would require a considerable number of new measurements to represent all conditions (e.g., farm sectors, animal productivity levels, management choices, climates) that could affect emissions. Regression analysis would be used to determine empirically which factors were of greatest importance. After having considered the emission factor approach, the committee recommended an alternative: to estimate emissions using process-based models rather than strictly empirical models.

Process-based models would involve analysis of the farm enterprise through study of its component parts. The analysis would use mathematical modeling and experimental data to simulate conversion and transfer of reactants and products through the farm enterprise. In many cases, this approach would make use of mass balance equations to represent mass flows and losses of major elements through the system. Development of process-based models would still require research studies to obtain measurements of transformations (changes in form of relevant compounds) and transfers (changes in location of compounds), as well as emissions in animal enterprises. Nonetheless, the estimates would be based on the understanding of processes inherent to animal production. System-level models would be needed to predict the effect of management changes simultaneously on different compounds emitted and to estimate the effects of management changes on multiple farms associated with animal production either directly or indirectly (e.g., crop production).

6

Government Regulations and Programs

INTRODUCTION

In recent decades, Congress has enacted numerous statutes to protect the environment. Often the detailed implementation of these federal laws is governed by regulations that are authorized by statute and promulgated by the responsible administrative agency. For most environmental laws, this is the U.S. Environmental Protection Agency (EPA), established in 1970. Other agencies also have environmental responsibilities—for example, the U.S. Department of Agriculture (USDA) for soil conservation and other agroenvironmental programs and the National Oceanic and Atmospheric Administration for coastal zones. Under some federal laws, including the Clean Air Act (CAA) and the Clean Water Act (CWA), states play an important role in implementing federal regulatory measures within their territories. In addition, states have enacted independent environmental statutes and regulations, which sometimes impose standards more stringent than those in federal law.

The federal Administrative Procedure Act (APA) governs the activities of federal agencies, including the EPA. Under the APA, most agency regulations are enacted by "notice and comment rulemaking." Under this procedure (5 USC § 553), the EPA publishes a notice in the *Federal Register* describing the terms or the substance of the proposed regulation; interested persons then have an opportunity to comment during a period of at least 30 days. Comments are normally submitted in writing, but oral presentations may also be permitted. After consideration of the material presented, the agency may revise and promulgate the regulation, accompanied by a "concise general statement" of its basis and purpose. When a statute requires that regulations be made "on the record after opportunity for an agency hearing," the agency must follow a more formal procedure, similar to an agency adjudication with a hearing and opportunity to cross-examine witnesses. Some environmental laws prescribe "hybrid rulemaking," which may require a public hearing (Findley, 2000).

Environmental statutes often require that regulations be enacted only after careful study and application of scientific principles. For example, CAA section 112 (42 USC § 7412) governs hazardous air pollutants (HAPs) and establishes the initial list of HAPs. The EPA administrator has the obligation to review the list periodically, in light of new scientific knowledge, and to revise it by rule. The administrator must add pollutants that threaten "adverse human health effects (including, but not limited to, substances which are known to be, or may reasonably be anticipated to be, carcinogenic, mutagenic, teratogenic, neurotoxic, which cause reproductive dysfunction, or which are acutely or chronically toxic) or adverse environmental effects whether through ambient concentrations, bioaccumulation, deposition, or otherwise . . . " (42 USC § 7412(b)(2)). Similarly, emission standards, established by regulation for categories of major sources and area sources of HAPs, must meet scientific standards (42 USC § 7412(d)). Application of good science in environmental regulation is essential to ensure both effective regulation and fairness to regulated entities as well as to withstand legal challenge.

Agriculture has long enjoyed favored status under the law, and agricultural operations have been exempt from numerous federal and state laws that govern other businesses. Environmental laws and regulations also often included "safe harbors" for agriculture (Ruhl, 2000). Despite the acknowledged impact of farming on the natural environment, agriculture has been "one of the last uncharted frontiers of environmental regulation" (Ruhl, 2000). Some laws are structured so that farms escape regulatory obligations; others exempt agriculture specifically from regulatory provisions (Ruhl, 2000). Therefore many of the federal and state laws that control air and water emissions from other sources have not governed agricultural activities.

Certain large animal feeding operations (AFOs), however, are subject to explicit environmental regulation under the Clean Water Act, and facilities that emit large quantities of air pollutants may be regulated under the Clean Air Act. For example, under the CWA, concentrated animal feeding operations (CAFOs; see Appendix B) are regulated as point sources of water pollution. Recent amendments to the Coastal Zone Management Act (CZMA) may impose management requirements on other livestock operations. Moreover, in recent years, large livestock operations that emit air pollutants such as ammonia (NH_3), hydrogen sulfide (H_2S), and particulates have been the focus of enforcement (or threat of enforcement) under the CAA. Some facilities may also be regulated under the Comprehensive Environmental Response, Compensation, and Liability Act (CERCLA, or the "Superfund" law) and the Emergency Planning and Community Right-to-Know Act (EPCRA), which require reporting when large quantities of pollutants are released. The material that follows provides a brief description of the federal laws that govern emissions from animal feeding operations.

Table 6-1 provides an overview of some of the provisions that may affect animal feeding operations under these federal statutes. More detailed information about regulated pollutants under the CAA appears in Appendixes G and H.

CLEAN AIR ACT

The Clean Air Act (42 USC §§ 7401-7671q), as amended, is the federal statute that governs air pollution. The CAA authorizes regulatory programs, including standards for ambient air quality to protect public health and welfare, special measures for regions that have not attained those standards, operating permits for stationary sources of air pollution, control technologies for new sources of air pollution, and measures to control hazardous air pollutants, as well as other programs (this chapter does not describe every CAA program). The CAA delegates rulemaking and enforcement authority to the federal EPA, which implements the act. States play an important role in carrying out CAA provisions and ensuring that state air quality meets federal air quality standards. States normally

TABLE 6-1 Overview of Federal Statutes and Their Provisions

Statute	Regulated Activity	Threshold	Impact
CAA	Major stationary source, 42 USC §§ 7602(j), 7661a	Emits or has potential to emit 100 tpy of any air pollutant	Permits, emission fees
	Major source of hazardous air pollutant, 42 USC § 7412	Emits or has potential to emit 10 tpy of a HAP or 25 tpy of HAPs	Emission standards, permits
CERCLA	Reportable releases, 42 USC § 9603	NH_3, H_2S: 100 ppd	Reporting requirement
EPCRA	Reportable releases, 42 USC § 11004	NH_3, H_2S: 100 ppd	Reporting requirement
CWA	NPDES point source requirements, § 33 USC 1342; 40 CFR part 122	Size or regulatory determination of CAFO status	NPDES permit (state or federal)
	EPA effluent limitations and performance standards, 40 CFR part 412	Size (1000 animal units)	Effluent limitation guidelines and performance standards in NPDES permit
CZMA	Nonpoint source pollution, 16 USC § 1455b	Large or small AFOs without NPDES permits	Management measures in coastal zones identified by states

NOTE: NPDES = National Pollutant Discharge Elimination System; ppd = pounds per day; tpy = tons per year.

have air pollution legislation consonant with the CAA, which authorizes state air pollution control agencies to implement the act in their territories through state implementation plans, permitting of air pollution sources, and other measures.

The CAA defines "air pollutant" as "any air pollution agent or combination of such agents, including any physical, chemical, biological, radioactive . . . substance or matter which is emitted into or otherwise enters the ambient air. Such term includes any precursors to the formation of any air pollutant . . ." (42 USC § 7602 (g)). In a practical sense, the "criteria pollutants" and the "hazardous air pollutants" are the major focus of regulation.

As one writer has noted, "[t]he centerpiece of the Clean Air Act has been the national ambient air quality standard (NAAQS) program" (Brownell, 2001). The CAA prescribes that "[n]ational primary ambient air quality standards . . . shall be ambient air quality standards the attainment and maintenance of which in the judgment of the [EPA] Administrator, based on such criteria and allowing an adequate margin of safety, are requisite to protect the public health" (42 USC § 7409(b)(1)). Secondary ambient air quality standards, when enacted, should be designed to protect the public welfare (42 USC § 7409(b)(2)). EPA has established primary NAAQS for the six "criteria pollutants" identified by EPA and regulated under the CAA (42 USC § 7409): sulfur dioxide (SO_2), nitrogen dioxide (NO_2), particulate matter (PM), carbon monoxide (CO), ozone (O_3), and lead (40 CFR part 50). These national ambient standards are then implemented by state implementation plans (SIPs) and new source performance standards (NSPSs) (Weinberg, 2000).

Hazardous air pollutants are pollutants that present a serious threat to human health or the environment. HAPs are identified in a statutory list (42 USC § 7412(b)) that can be modified by EPA regulation. EPA currently regulates 188 HAPs (EPA, 2002), and sources emitting HAPs are generally subject to a standard identified as MACT (maximum achievable control technology). Precursors of ozone (volatile organic compounds, or VOCs) and secondary PM2.5 (ammonia) are also considered air pollutants even though they are not listed as criteria pollutants or HAPs. A number of the air emissions produced by livestock facilities (see Appendix H) are pollutants regulated under the CAA.

Role of the EPA in Implementing the CAA

The U.S. Environmental Protection Agency is the federal agency responsible for implementing the Clean Air Act to "protect and enhance the quality of the Nation's air resources" (42 USC § 7401(b)). Among its responsibilities is promulgation of regulations to implement various programs set out in the CAA (42 USC § 7601; see 40 CFR parts 50, 51, 53, 55, 60, 61, 63, 70, 71). For example, by regulation the EPA has established NAAQS for criteria pollutants and MACTs for hazardous air pollutants from major pollution sources, and NSPSs for facilities that contribute significantly to air pollution. EPA is also responsible for designating nonattainment areas where air quality standards have not been met (42 USC §§ 7407, 7501-7502).

Part of EPA's responsibility includes oversight of state and local air pollution control agencies (often called state air pollution regulatory agencies, or SAPRAs). The EPA must approve state implementation plans for meeting federal NAAQS or, if it does not approve a state plan, must implement its own federal plan (42 USC §§ 7407, 7410). After EPA approval, SIPs have "the effect of federal law" (Weinberg, 2000). EPA also delegates to states the authority to issue operating permits for air pollution sources, as required under Title V of the 1990 Clean Air Act Amendments (the so-called Title V operating permit program, codified at 42 USC §§ 7661-7661f; 40 CFR parts 70, 71).

The Role of the States in Regulating Air Pollution

The CAA delegates a significant role in regulating air pollution from stationary sources to the states (mobile source standards are normally set by the federal government [Weinberg, 2000]), and local government agencies may also assume some responsibility. State legislation normally assigns this role to state or local air pollution control agency (see 42 USC § 7602(b), which defines air pollution control agency). For example, the Texas Commission on Environmental Quality serves as the state air pollution control agency in Texas; and some large metropolitan areas (e.g., Houston) also have local agencies. In California and Arizona, local air districts regulate air pollution. In the discussion that follows, the term SAPRA includes both state and local air pollution control agencies.

The CAA directs that each state has "primary responsibility for assuring air quality within the entire geographic area comprising such State by submitting an implementation plan for such State which will specify the manner in which national primary and secondary ambient air quality standards will be achieved and maintained" (42 USC § 7407(a)). State provisions must be at least as stringent as federal requirements (40 CFR part 51.101).

In addition to implementing federal CAA programs, many states have enacted additional air pollution provisions not required by federal law. For example, both Minnesota and Texas have state ambient air quality standards for hydrogen sulfide (Minn. Rules § 7009.0080; 30 TAC § 112.31-.34). In Minnesota, livestock production facilities are exempt from this standard while, and shortly after, manure is removed from barns or storage facilities (Minn. Stat. § 116.0713). In 1999, Colorado enacted "Regulation No. 2, Odor Emission," with special rules for housed commercial swine feeding operations. This regulation establishes odor standards and requires approved covers for "anaerobic process wastewater vessels and impoundments" to minimize emission of odorous gases (Colorado, 1999). Other states have taken different measures, including setback requirements for large livestock operations. These state programs, however, supplement the SIPs and other measures under the CAA.

The State Implementation Plans

A major responsibility of the state air pollution control agency is preparation and submission of the SIP, which must provide for the "implementation, maintenance, and enforcement" of the primary NAAQS standard in the state or an air quality control region in the state (42 USC § 7410(a)(1)). The SIP is a step in the translation of national ambient standards into emission limitations that will govern individual sources of air pollution. After approval by EPA, it can be enforced as both state and federal law.

The CAA articulates the basic content of SIPs, while regulations (40 CFR part 51) provide detailed requirements for the state plans. In general, the CAA provides that SIPs must be enacted after notice, public hearing, and local consultation; they must be revised to comply with federal regulatory changes, technical advancements, or EPA findings of inadequacy. Most importantly, SIPs must include enforceable emission limitations and other control measures (including economic incentives), as well as schedules for compliance. A program to enforce these measures must be included. The SIP must regulate the modification and construction of stationary sources and authorize an operating and construction permit program. Major stationary sources must be required to pay a permit fee. The SIP must also prohibit emissions that will cause unlawful interstate air pollution. It must provide for appropriate measurement of ambient air quality (including air quality monitoring, if prescribed by EPA) and may have to require sources to monitor emissions. The state must have adequate personnel, funding, and authority to carry out the SIP, as well as authority for emergency and contingency plans. The plan must also comply with other specific requirements of the CAA (42 USC § 7410(a)(2)). The SIP should satisfy CAA requirements for any nonattainment areas included in the plan territory so that these areas will comply with NAAQS. The CAA imposes additional requirements for SIPs in nonattainment areas (42 USC § 7502(c)).

Permits

The provisions of the SIP govern individual facilities through state permitting programs. Two state permitting programs generally apply: (1) the preconstruction permit and (2) the operating permit. The preconstruction permit is required under provisions that govern the prevention of significant deterioration (PSD) of air quality in areas where NAAQS have been met (42 USC § 7475) and provisions for nonattainment areas where NAAQS have not been met (42 USC § 7503). In both PSD and nonattainment areas, the preconstruction permit requirement applies to major new sources or major modifications of an existing source. The definition of "major" differs between PSD and nonattainment areas (40 CFR part 52.51(b)(1); Brownell, 2001).

In addition, under the CAA provision regulating SIPs, states are to include a program for "regulation of the modification and construction of any stationary source" to ensure that NAAQS will be achieved (42 USC § 7410(A(2)(C)). States enjoy considerable flexibility in this area, but the state permit requirement may also affect minor new or modified sources (Brownell, 2001).

The preconstruction permit will typically include, among other things, a description of proposed air pollution abatement systems and a determination of the allowable emission rate. Permit requirements must be particularly stringent in a nonattainment area. Under these CAA provisions, as implemented in the relevant SIP, someone who plans to construct a new AFO, feed mill, or cotton gin may have to obtain an air permit prior to construction.

The CAA now requires operating permits for stationary sources of air pollution (42 USC §§ 7661-7661f). Title V of the 1990 CAA Amendments added provisions that require states to develop a comprehensive program of operating permits for most "major sources" of air pollution. EPA has authority to approve each state's permit plan and each state-issued permit (Brownell, 2001). Permits include enforceable emission limitations and standards, a schedule of compliance, reporting requirements, and other conditions. The permit acts as a "shield" for the permittee, because permits may provide that facilities in compliance with the permit will be considered in compliance with "applicable provisions" of the CAA (42 USC § 7661c(f)).

Major sources, as defined by statute and EPA regulations, pay an annual permit fee based on total emissions of regulated pollutants. Fugitive emissions are not considered in determining whether a facility is a "major stationary source" of air pollution (42 USC § 7602(j); 40 CFR part 70.2). However, once the major source threshold (100 tons per year [tpy] of any pollutant) is met, the permit fee is determined by "actual emissions" of all regulated pollutants (40 CFR part 70.9), including fugitive as well as point source emissions. This may become a serious issue for ground-level area source (GLAS) PM emissions.

Because most agricultural operations are believed to be minor sources of air pollution, few agricultural facilities are required to comply with the operating permit requirement at present.

Enforcement

The CAA authorizes substantial penalties for violations of its provisions, including violations of the requirements imposed by permits (42 USC § 7413). The EPA administrator has authority to commence a civil action for an injunction or a civil penalty of up to $25,000 per day for each violation of certain CAA provisions. The CAA also authorizes administrative penalties, assessed after a formal administrative hearing, of up to $25,000 per day of violation (subject to a maximum of $200,000). Field citations issued for minor violations may include penalties with a maximum of $5,000 per day of violation. Moreover, knowing violation of certain CAA provisions may lead to criminal prosecution, which may result in a fine and/or imprisonment. The CAA also authorizes EPA to pay an award of up to $10,000 to an individual who provides information leading to either a civil penalty or a criminal conviction. States, too, have authority to enforce CAA provisions under EPA-approved state programs.

As a practical matter, enforcement against agricultural operations is often triggered by complaints when those operations are perceived to cause a nuisance. For example, the SAPRA in Texas (examples used in this section are based on the air pollution regulatory process used in Texas, other states may operate differently) may receive a complaint from a citizen alleging that an AFO is emitting a pollutant that interferes unreasonably with the complainant's enjoyment of his or her property (i.e., causes a nuisance such as odor). If the complaint is verified by SAPRA compliance personnel, the facility is issued a "notice of violation" (NOV) and is subject to a penalty or fine for violation of the state air pollution law or regulations. The NOV often leads to a determination that the facility must decrease its rate of pollution. The facility may have violated its permit conditions by emitting more than its allowable emission rate (AER). If so, the facility must comply with permit conditions and is subject to administrative penalties. If the facility was emitting at a rate equal to or less than its AER, the AER may be decreased often requiring more efficient and costly controls. The facility's permit may be amended to reflect the new AER and to require an improved air pollution abatement system.

Abatement Strategies

As a consequence of violations of state or federal standards, AFOs must respond to increasing pressures from their respective state air pollution control agencies to decrease pollutant emissions. The lack of science-based emission estimates for AFOs and other kinds of agricultural operations affects the regulatory process in several ways.

The state permitting process is designed to protect the public by ensuring that pollutant concentrations downwind from agricultural sources do not violate the ambient air quality standards (the NAAQS), which prescribe averaging time periods and maximum ambient concentrations for criteria pollutants. Estimates of downwind concentrations can be obtained with dispersion (Gaussian) models, given an emission rate. A proposed facility, with its associated abatement strategies, may demonstrate that it meets the NAAQS by dispersion modeling (40 CFR part 51.160 & App. W). As part of the preconstruction permitting process, if modeling of emissions demonstrates that a proposed facility will comply with the NAAQS, the facility can receive a permit that allows emission at the modeled emission rate (the allowable emission rate, or AER). However, emission rates are often determined from emission factors. If emission factors do not exist or are incorrect, the permitting process is flawed.

Emission factors are also used to develop emission inventories—that is, calculations of the annual masses of pollutants emitted by different sources. These emission inventories may be used in the regulatory process if, for example, an area does not attain NAAQS for a specific pollutant and the SAPRA must amend its SIP to improve air quality. These strategies usually involve decreases of emissions from all sources of that pollutant. Thus, an incorrect emission factor may mean that the attainment strategy specified in the SIP is not effective.

Particulate matter can be regulated as total suspended particulate (TSP)—that is, PM less than a nominal 40-μm aerodynamic equivalent diameter (AED), particulate less than 10 μm AED (PM10), and particulate less than 2.5 μm AED (PM2.5). The current PM10 and PM2.5 NAAQS are 150 and 65 $\mu g/m^3$, respectively, for 24-hour average concentrations (40 CFR parts 50.6, 50.7). Prior to 1987, the NAAQS for TSP was 260 $\mu g/m^3$, 24-hour concentration. A concentration at the property line that exceeds the NAAQS (determined by sampling and/or modeling) will result in violations of the respective state air pollution statute. When this occurs, the SAPRA may impose fines and decrease the AER established by permit. The facility will be required to install more efficient abatement systems to decrease its emission rate to comply with permit conditions.

Odors and Common-Law Nuisance Litigation

Because the Clean Air Act and its regulations generally rely on objective measures of pollutants, the regulatory process has not been effective in controlling odors, which are difficult to measure objectively (Grossman, 1994). Long before enactment of modern environmental statutes, common-law nuisance litigation was used to abate nuisances caused by pollution. Nuisance law has continued to be used to address odor problems from livestock facilities. Indeed, most agricultural nuisance cases have involved odor. A nuisance plaintiff may prevail if the conduct of the livestock operator interferes unreasonably with the use or enjoyment of property by other persons (private nuisance) or with the health, safety, and welfare of the public (public nuisance). Nuisance suits involve a process of judicial balancing, which considers factors that include the type of nuisance and the land use in the surrounding area. If successful, the suit may result in a court order that awards damages to the plaintiff or that forces the facility to close or change its practices (e.g., incorporate manure immediately) to minimize odor.

Since the late 1970s, the states have enacted right-to-farm laws, which significantly limit nuisance suits by protecting certain agricultural operations against nuisance claims. Individual state right-to-farm statutes vary, but most protect existing agricultural operations only from nuisance claims arising from a change in condition (e.g., new residential development) in the surrounding area. Some laws protect only facilities that comply with federal and state environmental standards or that pose no threat to public health and safety. Though one right-to-farm law

(*Bormann v. Board of Supervisors*, 584 N.W.2d 309 (Iowa) 1998; cert. denied sub nom. *Girres v. Bormann*, 525 U.S. 1172 (1999)) has been held unconstitutional, these laws mean that common-law nuisance actions are no longer always available to abate odor pollution from livestock facilities. Right-to-farm laws, however, do not prevent enforcement of federal and state environmental laws and regulations that govern livestock and other agricultural operations.

CERCLA AND EPCRA

The Comprehensive Environmental Response, Compensation, and Liability Act (42 USC §§ 9601-9675) authorizes programs to remediate uncontrolled or abandoned hazardous waste sites and assigns liability for the associated costs of cleanup when a responsible party can be identified. The Emergency Planning and Community Right-to-Know Act (42 USC §§ 11001-11050) establishes requirements for emergency planning and notification to communities about storage and release of hazardous and toxic chemicals. Both statutes have reporting requirements that may apply to large livestock facilities. CERCLA and EPCRA require reporting from facilities that release a "reportable quantity" of certain hazardous pollutants. The CERCLA definition of "release" (42 USC § 9601(22)(D)) excludes "the normal application of fertilizer." (Sweeten et al. [2000] wonder if the exclusion of normal application of fertilizer from the definition of "release" would also apply to "standard practices for application of manure or wastewater (spreading or irrigation)." An EPA document (EPA, 1998a) refers to normal application of fertilizers "in accordance with product instructions.") Under CERCLA section 103 (42 USC § 9603), the person in charge of a facility must notify the National Response Center of any release into the environment of a hazardous substance equal to or greater than the reportable quantity. Under EPCRA section 304 (42 USC § 11004(a)), a facility owner or operator must provide notice to state and local authorities of releases greater than the reportable quantity of substances deemed hazardous under CERCLA or extremely hazardous under EPCRA. AFOs do not appear in the list of sectors subject to EPCRA section 313, which governs the Toxics Release Inventory (EPA, 2000b). A "federally permitted release" (42 USC § 9601(10)) is excepted from these reporting requirements. Moreover, under CERCLA, a facility with "a continuous release, stable in quantity and rate" has less burdensome reporting obligations (42 USC § 9603(f)(2)).

The CERCLA definition of "hazardous substance" (42 USC § 9601(14)) triggers reporting under both CERCLA section 103 and EPCRA section 304. That definition includes, among other substances, hazardous air pollutants listed under the Clean Air Act (42 USC § 7412). Among the reportable substances released by livestock facilities are hydrogen sulfide, ammonia, and some volatile organic compounds (Sweeten et al., 2000). The reportable quantity for both ammonia and hydrogen sulfide is 100 pounds per day (ppd) (18.3 tpy) (40 CFR part 355, App.

A, listing extremely hazardous substances). Substantial penalties apply for failure to report (42 USC § 9603(b); 40 CFR part 355.50).

EPA has generally not enforced the reporting requirement against AFOs that release hazardous air pollutants, but CERCLA includes a broad citizen suit provision (42 USC § 9659) and EPCRA also allows citizen suits for some violations (42 USC § 11046(a)). Large livestock operations are therefore vulnerable to citizen suits for failure to comply with reporting under these statutes. In February 2002, the Sierra Club announced plans to sue a major poultry producer for violation of CERCLA reporting requirements for ammonia (ENS, 2002).

Recent discussion of the definition of the exempted federally permitted release may have implications for AFOs. Specifically, in an Interim Guidance published in 1999 (64 Fed. Reg. 71614), EPA suggested that the federally permitted release exemption would apply only when the release "is subject to a permit or control regulation under a CAA program that is specifically designed to control the hazardous substance or [extremely hazardous substance] release" (64 Fed. Reg. 71618). Thus, releases at facilities that are not subject to a permit or control regulation under the Clean Air Act or a SIP would not be considered "federally permitted releases." These facilities, including grandfathered facilities and minor sources, would be required to report releases or file a continuous release report under CERCLA and EPCRA (64 Fed. Reg. at 71617). The many animal feeding operations that operate without permits may be affected by this interpretation. Further, an independent regulatory requirement for volatile organic compounds as ozone precursors or for particulate matter would not trigger the federally permitted release exemption (64 Fed. Reg. 71618).

This definition proved controversial, in part because it could have subjected AFOs to reporting requirements under CERCLA and EPCRA. In June 2000, after review of comments and court challenges (EPA's notice of suspension of the Interim Guidance included the joint motion in *National Assoc. of Mfrs et al. v. Browner* (Nos. 00-1111, 00-1121, D.C. Cir., 2000) to hold court proceedings in abeyance until publication of the new Interim Guidance (65 Fed. Reg. 39615, 39616)), the EPA suspended the 1999 Interim Guidance (65 Fed. Reg. 39615). EPA finally published its new Guidance in April 2002 (67 Fed. Reg. 18899); at that time EPA also published a Guidance for CAA grandfathered sources (67 Fed. Reg. 19750). That Guidance, which focuses on emission sources regulated by the CAA, explains operation of the exemption for federally permitted releases under CERCLA and EPCRA. It operates as "a general guide to determine, on a case-by-case basis, whether an air release of a hazardous substance qualifies as a federally permitted release." These will generally be subject to a "relevant CAA permit or control regulation" (67 Fed. Reg. 18901). The Guidance addresses VOCs, PM, NO (nitric oxide), and NO_2 (in compliance with NO_x limits), but does not address ammonia specifically. For minor sources, with emissions below an annual regulatory threshold limit, and thus no applicable CAA permitting requirement, the Guidance indicates that releases during normal operations "in compliance with a federally enforceable threshold . . . would generally meet the definition of federally permitted releases in CERCLA . . . when the emission threshold limits or eliminates the release" (67 Fed. Reg. 18902-18903). Unanticipated releases, through incidents such as accidents or malfunctions, should be reported.

CLEAN WATER ACT

The Federal Water Pollution Control Act (33 USC §§ 1251-1387), significantly amended in 1972, is designed to "restore and maintain the chemical, physical, and biological integrity of the Nation's waters" (33 USC § 1251(a)). Known as the Clean Water Act, the law protects water quality by a combination of ambient water quality standards, limits on effluents, and permits.

The regulatory structure of the Clean Water Act distinguishes between point sources and nonpoint sources of water pollution. The National Pollutant Discharge Elimination System (NPDES) governs point sources, which may discharge pollutants only in compliance with a state (or federal) NPDES permit (33 USC § 1342). Pollution from nonpoint sources is governed by state water quality planning under sections 208 and 319 of the CWA (33 USC §§ 1288, 1313). Some nonpoint sources will be regulated through the CWA requirement that each state establish total maximum daily loads (TMDLs) for identified waters where effluent limitations are not stringent enough to meet water quality standards (33 USC § 1313(d)). A TMDL establishes the amount of a pollutant that an impaired water body can receive without exceeding water quality standards. Both point and nonpoint sources of the pollutant may be considered in establishing the TMDL (*Pronsolino v. Nastri*, 29 F.3d 1123 (9th Cir. 2002)).

AFOs Under Current CWA Regulations

Under the CWA and accompanying regulations, CAFOs are defined as point sources, subject to NPDES requirements (33 USC § 1362(14); 40 CFR part 122.23(a)). Since 1976, EPA regulations have provided more specific guidance. An "animal feeding operation" is a lot or facility where animals are confined and fed or maintained for a total of 45 days or more in any 12-month period, and where crops, vegetation, forage growth, or postharvest residues are not present during the normal growing season (40 CFR part 122.23(b)(1)). An AFO is a "concentrated animal feeding operation" if it meets size criteria or if it is designated as a CAFO on the basis of regulatory factors, including discharge of animal wastes and process wastewaters into waters of the United States (40 CFR part 122.23(c)). Using size criteria, an AFO is considered a CAFO if the operation confines more than 1000 EPA animal units (AUs; see Appendix B for the EPA and USDA definitions) or if it confines more than 300 EPA animal units and pollutants are discharged directly or through a man-made device. Under current regulations, no AFO is considered a CAFO if it discharges only in a 25-year, 24-hour storm event (40 CFR part 122, Appendix B).

EPA regulations (enacted in 1974) set effluent limitations and performance standards for feedlots with 1000 or more EPA AUs and with various types of animals and confinement configurations (40 CFR part 412). For these large operations, effluent limitation guidelines (ELGs) prescribe that there shall be no discharge of process wastewater pollutants (water used in the feedlot that comes into contact with manure, litter, bedding, or other material used in production) to navigable waters. An exception allows discharge caused by chronic or catastrophic rainfall, when the facility is designed, constructed, and operated to contain all process-generated wastewaters plus runoff from a 25-year, 24-hour rainfall event (40 CFR part 412.13). The 25-year, 24-hour standard applies to facilities using the best available technology (economically achievable) (BAT). For facilities using the best practicable control technology (currently available) (BPT), the exception applies if the facility can contain process-generated wastewaters plus runoff from a 10-year, 24-hour rainfall event (40 CFR part 412.12). Performance standards for new sources prohibit discharge of process wastewater pollutants to navigable waters, with an exception for a 25-year, 24-hour rainfall event (40 CFR part 412.15).

EPA regulations governing both ELGs and NPDES permits refer to the 25-year, 24-hour rainfall event. The amount of precipitation that constitutes a 25-year, 24-hour rainfall event varies by location. In its ELGs, the EPA definition is "a rainfall event with a probable recurrence interval of once in . . . twenty-five years, . . . as defined by the National Weather Service in Technical Paper Number 40, 'Rainfall Frequency Atlas of the United States', May 1961, and subsequent amendments, or equivalent regional or state rainfall probability information developed therefrom" (40 CFR part 412.11(e)). For example, in Midwestern states (Illinois, Indiana, Iowa, Kentucky, Michigan, Minnesota, Missouri, Ohio, and Wisconsin), the 25-year, 24-hour rainfall amounts ranged from 3.5 to 7 inches (Huff and Angel, 1992).

NPDES permits govern the discharge of pollutants by CAFOs, and compliance with permit requirements normally constitutes compliance with requirements of the CWA (33 USC § 1342 (a), (k)). NPDES permits can impose both technology-based effluent limitations (for CAFOs, the ELGs) and limitations based on water quality. Though permits normally incorporate the ELGs, regulators use their best professional judgment to impose stricter limitations if water quality standards cannot be met through ELGs. CAFOs that are not large enough to be covered by ELGs (those with fewer than 1000 EPA animal units) are subject to limits determined by best professional judgment (EPA, 2000c).

The CWA allows states to assume responsibility for implementation of the NPDES program, provided that their NPDES requirements are at least as strict and as broad as federal requirements. At the end of 2000, 43 states and the Virgin Islands had authority to implement the NPDES program (66 Fed. Reg. 2960, at

2964). Oklahoma, which has other NPDES program authority, has no authority to regulate CAFOs. In the remaining states (Alaska, Arizona, Idaho, Maine, Massachusetts, New Hampshire, New Mexico, and the District of Columbia), the appropriate EPA regional office issues NPDES permits. CAFO permits, like other NPDES permits, may be individual permits, designed for a single facility, or general permits, designed for a category of facilities. Once a general permit is approved, individual facility operators file a notice of intent to be covered by the general permit (66 Fed. Reg. 2960, at 2964-65). In March 2001, the EPA published a report *State Compendium: Programs and Regulatory Activities Related to Animal Feeding Operations* (EPA, 2001d), which describes the various state programs (see 66 Fed. Reg. at 2968-70 for a summary of state NPDES implementation and other regulation of CAFOs).

Under current law and regulations, a relatively small number of AFOs are required to obtain NPDES permits. EPA estimates that about 12,660 CAFOs confine more than 1000 EPA animal units, but that only about 4000 CAFOs have NPDES permits (66 Fed. Reg. 2960, at 2968-69; EPA, 2001b). AFOs that are not regulated as point sources are considered nonpoint sources, which are subject to rather weak state planning programs under the CWA.

Proposed Regulatory Changes

Current measures that regulate CAFOs date from the 1970s. By the early 1990s, regulators had concluded that consolidation and other changes in the livestock industry required reconsideration of CAFO regulation (66 Fed. Reg. 2965). In the late 1990s, both pork producers and the poultry industry worked with EPA to develop voluntary environmental compliance programs (66 Fed. Reg. 2966).

In 1998, the *Clean Water Action Plan* mentioned polluted runoff from agriculture as one of the serious water quality problems facing the United States (EPA and USDA, 1998). The plan recommended both that EPA publish and implement "an AFO strategy for important and necessary EPA actions on standards and permits" and that EPA and USDA "jointly develop a unified national strategy to minimize the environmental and public health impacts of AFOs." EPA published two documents, the *Draft Strategy for Addressing Environmental and Public Health Impacts from Animal Feeding Operations* (EPA, 1998b) and the *Compliance Assurance Implementation Plan for Animal Feeding Operations* (EPA, 1998c). USDA and EPA (1999) cooperated on a *Unified National Strategy for Animal Feeding Operations*, which establishes a national goal (to minimize water pollution from confinement facilities and land application of manure) and performance expectations (Comprehensive Nutrient Management Plans, CNMPs) for AFOs.

In January 2001, EPA published its proposed new CAFO regulations (66 Fed. Reg. 2960) that would revise both the NPDES regulations defining CAFOs and requiring permits (40 CFR part 122) and the ELGs that set technology-based standards for beef, dairy, swine, and poultry CAFOs (40 CFR part 412). Proposed amendments are expected to increase the number of facilities that will be defined as CAFOs and must therefore operate under an NPDES permit. The following paragraphs highlight some of the proposed amendments. Proposed amendments have been published, with extensive commentary and analysis, (66 Fed. Reg. 2960). A summary of proposed changes appears in EPA (2001b). A clearly organized, detailed summary of proposed changes, in tabular form, can be found in EPA (2001c).

Changes proposed for the NPDES CAFO regulations (40 CFR part 122) include a new definition of animal feeding operation, which would distinguish AFOs (with stabled or confined animals) from operations with animals on pasture or rangeland. AFOs would include both the production area and the land application area. Proposed regulations would amend the definition of CAFO; the EPA is considering a two-tiered structure (with the CAFO threshold set at 500 AUs) or a three-tiered structure, which would require mid-tier facilities (300 to 1000 AUs) to certify that they are not a CAFO or to obtain a permit. The proposal would eliminate the exemption providing that an AFO that discharges only in a 25-year, 24-hour storm event is not a CAFO. All CAFOs would be obligated to apply for a permit. New animal and facility types would be regulated; these include poultry operations with dry manure handling, stand-alone swine nurseries and heifer operations, and veal operations. Further, the land application area would be included in the CAFO definition, and each CAFO would be required to prepare and implement a Permit Nutrient Plan to govern land application at agronomic rates. The agricultural storm water exemption (from the CWA, 33 USC § 1362(14)) would pertain only when manure is applied under "proper agricultural practices." EPA is considering other approaches to prevent CAFOs from using the agricultural storm water exemption. Some off-site recipients of CAFO manure would also face regulation (66 Fed. Reg. 2960; EPA, 2001b).

Permitting requirements would be changed. For example, processors that exercise "substantial operational control" over contract growers would be required to apply for a permit, either alone or with other owners or operators. Co-permitting could be waived in states with effective programs for excess manure. CAFOs would be required to maintain a permit until the facility and its manure storage were properly closed. Additional permit requirements may apply if a CAFO is in an area where groundwater has a direct hydrologic connection with surface waters (66 Fed. Reg. 2960; EPA, 2001b).

The proposed regulations also would change the effluent guidelines for CAFOs (40 CFR part 412). ELGs would apply to all beef, dairy, swine, veal, and poultry facilities that meet the new definition of CAFO under the amended NPDES regulation. Beef and dairy CAFOs and new swine, poultry, and veal CAFOs would have to determine whether a hydrologic link exists between groundwater (under the feedlot and manure storage areas) and surface waters. No discharges would be permitted from swine, veal, and poultry CAFOs, nor would an overflow allowance be permitted. Routine inspection of production areas would be required, and open liquid impoundments must have depth markers. CAFOs must handle dead animals in a way that does not pollute waters (66 Fed. Reg. 2960; EPA, 2001b)

Other proposed ELG regulations would govern land application of manure. CAFOs would be required to prepare and implement a Permit Nutrient Plan (PNP) and to apply manure on the basis of crop nutrient requirements. The land application rate may be calculated based on the phosphorus index, the phosphorus threshold, or on phosphorus determined by soil test. Detailed record keeping, as well as manure and soil sampling, would be required to prove compliance with the PNP. Set-back requirements would prohibit application of manure and wastewater within 100 feet of surface water (66 Fed. Reg. 2960; EPA, 2001b).

In November 2001, EPA published its Notice of Data Availability (NODA) on the proposed rule (66 Fed. Reg. 58556), a detailed document outlining comments and data received and describing how the data may be used in the final CAFO regulations. EPA requested comments on a number of issues. For example, the agency asked for comments about the possible use of the environmental management system ("a continual cycle of planning, implementing, reviewing, and improving the actions an organization takes to meet its environmental obligations" (66 Fed. Reg. 58601)), as a way to give states flexibility in managing CAFO programs. For proposed ELG requirements, among other issues, EPA asked for comments about technical feasibility, costs, and benefits of zero-discharge standards for swine and poultry operations and about reasonable amounts of phosphorus banking as an acceptable nutrient management practice. For proposed NPDES requirements, EPA asked for comments on an alternative three-tier structure for defining CAFOs, using 500 EPA AUs, instead of 300, to define the middle tier; size thresholds for dry-lot duck operations; options for defining horse operations as CAFOs; and other issues. The NODA includes a brief consideration of air emissions from CAFOs (66 Fed. Reg. 58592-58593).

In July 2002, the EPA published a second NODA (67 Fed. Reg. 48099). That notice presented information and requested comment on alternative regulatory thresholds for chicken operations using dry litter management, the possible creation of alternative performance standards for CAFOs, and possible refinements in the EPA economic analysis model. For chicken operations, EPA is considering whether the 1000 EPA AU equivalent for broilers and laying hens should be changed to 125,000 and 82,000 birds, respectively, to reflect manure generation rates more accurately. EPA asked for comments on these alternative EPA AU equivalents.

EPA is also considering, and has solicited comments on, a possible framework for alternative performance standards that would encourage voluntary development and implementation of effective technologies and management practices. The Production Area Approach would involve performance standards to govern manure and wastewater discharges; CAFOs could discharge treated process wastes if treatment would result in pollution control equivalent to or better than BAT standards. Under the whole farm approach, CAFOs would use an audit process to evaluate and implement improvements on the whole farm, including land application areas, and would have a discharge allowance for the production area. The plan developed under this approach should result in improvement across "multiple environmental media," including air emissions. The NODA solicits comments about several aspects of the proposed alternative standards.

Finally, EPA explained possible changes to its model framework and assumptions and to the baseline financial data used to assess the economic effects of its final regulations on CAFOs. EPA is considering these changes as a result of comments on its earlier proposal as well as new data.

COASTAL ZONE MANAGEMENT ACT

The Coastal Zone Management Act of 1972 (16 USC §§ 1451-1465) is designed to protect natural systems in the U.S. coastal zone in the face of competing demands that threaten ecological and other values. The act authorizes federal grants to coastal states that protect coastal land and water by developing and implementing management programs consistent with the CZMA, which defines the "coastal zone" to include coastal waters and adjacent shorelands. This zone extends inland "only to the extent necessary to control shorelands, the uses of which have a direct and significant impact on the coastal waters" (16 USC § 1453). States may include entire watersheds where land uses have a "direct and significant" impact on coastal waters (15 CFR part 923.31). States enjoy discretion in delineating the borders of their coastal zones (Malone, 2001, § 2.5). Because livestock facilities exist in close proximity to coastal waters (e.g., in North Carolina), state CZMA programs have implications for AFOs, particularly after 1990 amendments to the law.

The Coastal Zone Act Reauthorization Amendments (CZARA) added a new section to the CZMA (16 USC § 1455b). Acknowledging the importance of nonpoint source pollution in degrading coastal waters, the new section requires states with federally approved coastal zone management programs to develop and implement management measures for nonpoint source pollution. The new section required EPA to provide guidance for specifying management measures for sources of nonpoint pollution in coastal waters. EPA addressed agricultural runoff in its *Guidance Specifying Management Measures for Sources of Nonpoint Pollution in Coastal Waters* (EPA, 1993b). In this document, EPA specified management measures for "confined animal facilities" that affect coastal waters. Guidelines define large and small confined animal facilities, using thresholds lower than CAFO regulations. A large beef facility, for example, has 300 head (EPA animal units) or more; a small facility has 50-299 head (EPA animal units). Operations that are CAFOs under the Clean Water Act and therefore have NPDES permits are not subject to CZARA management measures (66 Fed. Reg. 2960, at

2968; EPA, 1993b). The CZARA guidance for confined animal facilities prescribes waste discharge limits and standards for waste storage structures. The guidance also includes nutrient management measures (including a nutrient management plan) and recommendations for grazing animals.

THE ROLE OF USDA

Like EPA, USDA is responsible for implementing federal statutory programs, especially programs authorized in federal agricultural legislation (the so-called Farm Bills). The USDA, founded in 1862 and last reorganized in 1994, is authorized to manage a diverse range of programs. Including food programs (e.g., food stamps, school lunch), management of national forests, rural development, safety of meat and poultry products, agricultural trade, and conservation of natural resources. Several USDA agencies have conservation responsibilities that may involve livestock and their environmental effects. For example, the Natural Resources Conservation Service (NRCS) plays an important role in implementing conservation programs; the Farm Service Agency (FSA) also implements conservation programs. State and county field offices play a role in implementing USDA programs at the local level. The Agricultural Research Service (ARS) and the Cooperative State Research, Education, and Extension Service (CSREES) have research and educational responsibilities. Some of the responsibilities of these agencies are discussed below.

USDA cooperates with EPA when issues concern both agriculture and environment. For example, USDA (NRCS) and EPA have collaborated on the *Unified National Strategy for Animal Feeding Operations* (USDA and EPA, 1999; discussed above) and other matters related to the measures governing CAFOs and the rules for TMDLs.

Some USDA Operating Programs

NRCS

NRCS was established in 1994, as part of the USDA reorganization, to carry out certain USDA functions related to natural resources and the environment. Through technical personnel assigned to field offices, NRCS provides technical assistance and information, as well as financial assistance, to landowners, agricultural producers, and others. It helps individuals to implement conservation systems and practices (sometimes with cost-share funding), often in cooperation with local conservation districts; millions of acres are protected annually through NRCS efforts. NRCS also helps government units and community groups to protect the environment through resource planning, including farmland and watershed protection. The agency conducts inventories and assessments of U.S. natural resources (e.g., soil surveys) and makes this information available to the public.

NRCS also develops and maintains technical, science-based standards for conservation. NRCS assists other USDA agencies; for example, it provides technical assistance for FSA implementation of the Conservation Reserve Program and the conservation compliance requirements of other farm programs (NRCS, 2002c).

NRCS manages voluntary conservation programs authorized by federal Farm Bills beginning in 1985. These include the Environmental Quality Incentives Program (discussed in detail below), Conservation of Private Grazing Land Program, the Conservation Security Program (to be operative in fiscal yar 2003), Farmland Protection Program, Wetlands Reserve Program, Wildlife Habitat Incentives Programs, and others. Funding for several of these programs comes from the Commodity Credit Corporation (CCC).

NRCS personnel work with owners and operators of AFOs, with the objective of helping them to develop and implement Comprehensive Nutrient Management Plans (NRCS, 2002a). In 2001, for example, NRCS helped producers to apply nutrient management systems on 5.4 million acres and "[p]lanned or applied 10,500 waste management systems, including waste storage structures, treatment lagoons, composting facilities, and roof runoff management" (NRCS, 2001). NRCS has prepared a Draft Comprehensive Nutrient Management Planning Technical Guidance, as part of its *National Planning Procedures Handbook* (NRCS, 2002d). The Guidance is intended for those who develop, or assist others to develop, CNMPs.

FSA

The Farm Service Agency, also established in 1994 has been given primary responsibility for several important USDA programs. These include farm commodity programs; disaster assistance; farm ownership, operating, and emergency loans; and food aid programs. In addition, FSA manages several USDA resource conservation programs. FSA, like NRCS, operates through its field offices, which include state offices and 2500 USDA service centers (FSA, 2002).

FSA has primary responsibility, with NRCS assistance, for the Conservation Reserve Program (CRP). The CRP is a voluntary program under which eligible land (e.g., highly erodible cropland, marginal pasture land) is enrolled by contract. In exchange for an agreement to take the land out of production and employ approved conservation practices for a 10-15 year period, owners and operators receive annual rental payments and cost-share assistance to establish approved conservation practices on the land. Farm legislation enacted in 2002 increased the maximum CRP enrollment from 36.4 million to 39.2 million acres (P.L. 107-171, § 2101, codified at 16 USC § 3831). FSA also implements the Conservation Reserve Enhancement Program, a federal-state partnership, and the Emergency Conservation Program, which provides cost-share payments to producers for the rehabilitation of farmland damaged by natural disasters.

The 2002 Farm Bill and Livestock Producers

The Farm Security and Rural Investment Act of 2002 (the 2002 Farm Bill) (P.L. 107-171) includes a Conservation Title, which makes financial and other assistance available to livestock producers, as well as other farmers. The Conservation Title reauthorized and amended a number of conservation programs enacted in prior agricultural legislation. Particularly relevant is the Environmental Quality Incentives Program (EQIP), part of a program now called the Comprehensive Conservation Enhancement Program (under prior law, the Environmental Conservation Acreage Reserve Program). EQIP was created in 1996, in part to help livestock and other producers comply with federal and state environmental regulations. Under the 2002 Farm Bill, EQIP was reauthorized through 2007, and authorized funding was increased significantly (to $1.3 billion in fiscal year 2007). USDA's Natural Resources Conservation Service administers EQIP, with funding from the Commodity Credit Corporation (67 Fed. Reg. 48431).

EQIP is intended "to promote agricultural production and environmental quality as compatible goals, and to optimize environmental benefits." Among other purposes, the program helps producers to comply with regulatory requirements concerning soil, water and air quality, wildlife habitat, and surface and groundwater conservation. The focus on air quality was added in 2002; prior law referred to "soil, water, and related natural resources," with no specific mention of air quality (16 USC § 3839aa, before amendment by P.L. 107-171). In a July 2002 notice, the CCC announced its intention to issue a proposed rule for fiscal years 2003 through 2007. Among the issues to be considered in the proposed rule is "integration of air quality as a program goal" (67 Fed. Reg. 48431). The notice listed changes that the 2002 Farm Bill required for fiscal year 2002 administration of EQIP. EQIP can help producers "to make beneficial, cost effective changes to . . . nutrient management associated with livestock." The program is also intended to assist producers in meeting environmental quality criteria, to provide assistance to install and maintain conservation practices and to help streamline conservation planning and regulatory compliance (Farm Bill § 2301, codified at 16 USC § 3839aa).

To carry out these purposes, EQIP authorizes contracts, lasting from 1 to 10 years, with producers who agree to implement eligible environmental and conservation practices in exchange for cost-share and incentive payments, as well as technical assistance. "Practice" is defined to include structural practices, land management practices, and comprehensive nutrient management planning practices (16 USC § 3839aa-1(5)). A livestock producer who develops a CNMP is eligible for incentive payments. In determining the amount and rate of incentive payments, "great significance" can be given to a practice that promotes "residue, *nutrient*, pest, invasive species, or *air quality management*" (16 USC § 3839aa-2(a), (e), italics added). Payments to an individual or entity are limited to $450,000 for all contracts entered during fiscal years 2002 through 2007. Beginning in fiscal year 2003, however, EQIP payments may not be made to an individual or entity whose average adjusted gross income for the previous three years exceeds $2.5 million, unless 75 percent of that income came from farming, ranching, or forestry (NRCS, 2002b).

The 2002 Farm Bill makes more EQIP money available to livestock producers. Under prior law, livestock producers were to receive at least 50 percent of EQIP funding, but the 2002 Farm Bill targets 60 percent of program funding for environmental practices relating to livestock production (§ 3839aa-2(g)). Amendments to the program eliminate the preference that participants be located in a specially designated conservation priority area and extend eligibility to livestock producers throughout the United States. Moreover, an animal unit cap no longer limits eligibility for cost-share payments for constructing animal waste management facilities. Applications from producers who operate confined livestock feeding operations must provide for development and implementation of a comprehensive nutrient management plan (§ 3839aa-5(a)(3)).

In contrast to EQIP, federal money is not available for animal waste facilities under the Conservation Security Program (CSP), enacted in the 2002 Farm Bill. The CSP pays producers for adopting or maintaining conservation practices that help to protect or improve the quality of soil, water, air, energy, and plant and animal life and other conservation purposes. The CSP includes three tiers of conservation contracts, with increasingly stringent requirements. Eligible producers enter conservation contracts that set out the required conservation practices; a variety of practices are eligible. In exchange, producers receive payment and a share (normally 75 percent; 90 percent for a beginning farmer) of the cost of adopting or maintaining the required conservation practices. Livestock farmers are not excluded from the CSP, but the statute specifies that payment may not be made for "construction or maintenance of animal waste storage or treatment facilities or associated waste transport or transfer devices for animal feeding operations" (Farm Bill § 2001, codified at 16 USC § 3838c(b)(3)(A)). Regulations for the CSP have not been enacted, and the program will not be effective until some time during 2003.

USDA Research and Extension

ARS

The Agricultural Research Service, established in 1953, is the in-house research agency of USDA. Its responsibilities are articulated by statute. In brief, federally supported agricultural research, extension, and education are intended to enhance the competitiveness of the U.S. agriculture and food industry; increase long-term productivity, while protecting the natural resource base; develop new uses and products for agricultural commodities; promote economic opportunity in rural communities and meet U.S. information and technology transfer needs; improve risk management; improve safe production and processing of food and fiber resources, using methods that balance yield and environmental soundness;

support higher education in agriculture; and maintain the food supply (7 USC § 3101).

ARS research, carried out by field scientists located throughout the United States, involves more than 1200 projects organized into 22 national programs. Among those related to livestock production are the national programs directed to air quality and manure and by-product utilization. Research components of the air quality program focus on particulates, agriculturally emitted ammonia, and malodorous compounds from animal production operations, as well as ozone impacts and volatilized pesticides and other synthetic organic compounds. Among the projected outcomes related to livestock facilities are "[i]mproved understanding of the physics of dust emissions . . . with state-of-knowledge control measures," a "database of ammonia emission factors for animal production facilities, manure storage areas, and manure and fertilizer field application sites," and improved understanding of the formation, interaction, and transport of odorous compounds (ARS, 2002). The ARS National Program on manure and by-product utilization focuses on nutrient management (protection of soil, water, and air from excess nutrients), atmospheric emissions, and pathogens. Projected outcomes include more efficient conversion of feed, identification of alternative uses of manure, development of "management practices, treatment technologies and decision tools" to improve CNMPs and help meet TMDLs, and practices and technologies to control pathogens (ARS, 2002).

CSREES

The Cooperative State Research, Education, and Extension Service carries out USDA functions related to cooperative state research programs and cooperative extension and education programs (7 USC § 6971). The mission of CSREES is to advance "creative and integrated research, education, and extension programs in the food, agriculture and related sciences" (CSREES, 2002). Among its primary functions are leadership for programs that support university-based and other institutional research, education, and extension, and administration of federal assistance for these programs. To accomplish its mission, CSREES works with the land-grant universities, other colleges, universities, and research and educational organizations to develop programs for agricultural research, extension, and higher education. Land-grant universities and other partners carry out the programs (CSREES, 2002). Land-grant universities and certain other educational institutions receive funds, allocated to states by formula, to carry out research and outreach programs in food and agricultural sciences.

CSREES, like ARS, has projects related to livestock production. For example, an animal waste management program, with nationwide collaboration, is intended to decrease environmental impairment and achieve other environmental benefits by educating producers, increasing the use of best management practices, providing training for CAFOs, and other activities (CSREES, 2001).

SUMMARY

Many different federal environmental statutes and their associated regulations may affect AFOs. Air pollutants are regulated primarily by the Clean Air Act, which includes measures to govern criteria pollutants and hazardous air pollutants. States play an important role in implementing CAA provisions and issuing permits for facilities under their state implementation plans. Some air emissions are also regulated by CERCLA and EPCRA. Water pollution is governed by the Clean Water Act and, in some areas, the Coastal Zone Management Act. The EPA plays a major role in implementing these statutes, with significant cooperation from the states. In addition to these regulatory programs, statutory programs implemented by the USDA offer technical assistance and financial incentives to livestock producers.

7

Improving Knowledge and Practices

INTRODUCTION

The committee was asked by the U.S. Environmental Protection Agency (EPA) and the U.S. Department of Agriculture (USDA) to develop both short-term (5 years or less) and long-term (20-30 years) research recommendations. The charge to provide recommendations on science-based methodologies and modeling approaches for estimating animal feeding operation (AFO) air emissions indicates a desire by EPA to limit those emissions or by USDA to quantify and evaluate reductions of emissions for specific air pollution abatement strategies or management practices. Air pollution regulations have generally been based on emission inventories (estimated total annual emissions from various sources), dispersion modeling, and concentration measurements, followed by permitting and enforcement. If a specific operation were to be found responsible for annual emissions adversely affecting regional ozone, for example, the relevant emissions from that operation might have to be reduced.

Two possible approaches were considered by the committee for providing recommendations on the most promising science-based methodologies and modeling approaches for estimating and measuring emissions: (1) to base recommendations on the current regulatory approach, which the committee refers to as the "emissions factor" approach to characterize air emissions from representative AFOs; and (2) to use a "process-based" model of individual AFOs to estimate the flow of emission-generating substances through the sequential processes of the farm enterprise.

The emissions factor approach uses research-based estimates of the relationship between estimated emissions of various kinds and some readily estimated base (e.g., number of animals, acres, volume of manure) as a multiplier to obtain estimates of total emissions. The process-based approach for estimating air emissions is favored by the committee for most kinds of emissions as the primary focus for both the short- and the long-term research recommendations. In some cases where a process-based model may not be feasible or appropriate, at least until further research has been done, (i.e., particulate matter [PM] and volatile organic compounds [VOCs] as the main constituents of odor), the research recommended to improve emission estimates allows for an emissions factor approach.

This section of the report outlines short- and long-term research programs to achieve not only reductions in atmospheric emissions of the substances of concern but also of all losses from AFOs. While the report focuses on specific species and elements, an important aspect of AFOs is that they are biogeochemical systems, and as such there are significant interactions among elemental cycles. Just as the committee recommends that controls on specific substances (e.g., ammonia [NH_3]) need to be done on a total system approach (e.g., all N-containing substances), controls on individual elements should be designed in a biogeochemical context.

Because of their direct regulatory and management responsibilities for mitigating the effects of air emissions, EPA and USDA should be expected to provide substantial resources to support both short-term and long-term research programs on air emissions. The fact that USDA has by far the largest overall research program might suggest that it provides the bulk of the needed research funds. A change in research priorities in both agencies is needed if air emissions are to be addressed with an adequate base of scientific information.

FINDING 12. USDA and EPA have not devoted the necessary financial or technical resources to estimate air emissions from AFOs and develop mitigation technologies. Scientific knowledge needed to guide regulatory and management actions requires close cooperation between the major federal agencies (EPA, USDA), the states, industry and environmental interests, and the research community, including universities.

RECOMMENDATIONS:

- **EPA and USDA should cooperate in forming a continuing research coordinating council: (1) to develop a national research agenda on issues related to air emissions from AFOs in the context of animal production systems, and (2) to provide continuing oversight on the implementation of this agenda. This council should include representatives of EPA and USDA, the research community, and other relevant interests. It should have authority to advise on research priorities and funding.**

- **Exchanges of personnel among the relevant agencies should be promoted to encourage efficient use of personnel, broadened understanding of the issues, and enhanced cooperation among the agencies.**
- **For the short term, USDA and EPA should initiate and conduct a coordinated research program designed to produce a scientifically sound basis for measuring and estimating air emissions from AFOs on local, regional, and national scales.**
- **For the long term, USDA, EPA, and other relevant organizations should conduct coordinated research to determine which emissions (to water and air) from animal production systems are most harmful to the environment and human health and to develop technologies to decrease their releases into the environment. The overall research program should include research to optimize inputs to AFOs, optimize recycling of materials, and significantly decrease releases to the environment.**

FINDING 13. Setting priorities for both short- and long-term research on estimating air emission rates, concentrations, and dispersion requires weighing the potential severity of adverse impacts, the extent of current scientific knowledge about them, the potential for advancing scientific knowledge, and the potential for developing successful mitigation and control strategies.

RECOMMENDATIONS:

- **Short-term research priorities should improve estimates of emissions on individual AFOs including effects of different control technologies:**
  1. **Priority research for emissions important on a local scale should be conducted on odor, PM, and hydrogen sulfide (H_2S) (also see Finding 2).**
  2. **Priority research for emissions important on regional, national, and global scales should be conducted on ammonia, nitrous oxide (N_2O), and methane (also see Finding 2).**
- **Long-term research priorities should improve understanding of animal production systems and lead to development of new control technologies.**

SHORT-TERM RESEARCH

As stated earlier in this report, some of the committee's findings provide a basis for organizing recommendations for short-term research needs and actions. These recommendations are intended to guide research that will provide EPA and USDA with information that will help direct regulatory and management actions within the next five years. The recommended research directions could be extended beyond the five-year period if the research is not adequately funded for completion within that time, or as new needs and directions become apparent.

Some of the research proposed in the five-year program is an extension of research already initiated to support the current emission factor approach. However, some—indeed much—of the recommended research in the five-year program is needed to begin implementing the committee's proposed process-based modeling approach.

The relative importance of local, regional, and global impacts of AFO emissions determines research priorities (see Finding 2). The primary emissions of local concern are PM and odor (including VOCs that contribute to odor). The overall goal of the research is to provide information that can help decrease the emissions of PM and odors to minimize impacts on the public near the sources of the emissions. At the global, national, and regional scales, the greenhouse gases methane (CH_4) and nitrous oxide (N_2O), along with ammonia are the emissions of major concern. Ammonia, NO (nitric oxide), and VOCs from AFOs are precursors of secondary PM2.5 and ozone, and may have impacts on a regional scale.

For local impacts, ambient concentrations near the "fence line" or the nearest occupied residence are important because they may be associated with health effects and are now used in the regulatory process.

Permitting, dispersion modeling, source sampling of emission points, and ambient concentration measurements near the source are used as a basis for establishing the emission rate limits in permits. The current EPA goal is to set emission rate limits at amounts that will not exceed National Ambient Air Quality Standards (NAAQS) for criteria pollutants or screening concentrations for hazardous air pollutants (HAPs). Directly measuring emissions from every AFO poses both serious technical problems and prohibitive costs. Thus, there is a need for an approach that can be used by state and federal agencies to estimate emissions from individual AFOs.

The emission factor approach has been used successfully for the nonagricultural sector. It is only recently that some of the agricultural emission factors needed by state and federal Air Pollution Regulatory Process (APRP) agencies have been measured. In many cases, the necessary factors either do not exist or are inaccurate. The committee believes that its recommendation for process-based modeling to estimate the flow of emission-generating substances through the sequential processes of the farm enterprise is a better means to measure or estimate the quantities of AFO emissions from particular operations.

Short-Term Research Recommendations

This section includes recommended research needs that address both the science basis of the process-based approach in the short term (up to five years) and the continued improvement of this approach. In addition, critical short-term emission research for PM and odor-related activities are recommended.

NEED 1. Scientifically sound and practical protocols for measuring air concentrations, emission rates, and fates are needed for the various elements (nitrogen, carbon, sulfur), compounds (e.g., NH_3, CH_4, H_2S), and particulate matter (Finding 7).

Accurate concentration measurements are needed to estimate the off-property impacts of AFO emissions. Chapter 3 addresses concentration measurements for specific emissions. Different methods for measuring concentrations may be used for source sampling in the exhaust stream and in the ambient air. Two concerns are relevant when comparing protocols for emissions from AFOs with those from other sources:

- PM emissions from agricultural operations characteristically have larger particle sizes than those associated with urban areas or stack emissions from industrial sources. Buser et al. (2002) documented that sampling PM10 or PM2.5 with EPA recommended and approved samplers will result in significant errors for PM with a mass median diameter (MMD) larger than 10 μm aerodynamic equivalent diameter (AED). (The error increases with an increase in MMD.)
- Emission rates for some nonagricultural sources (e.g., chemical or power plants) tend to be more regular in any 24-hour period and less dependent on season and weather than those for AFOs. For these sources, it is common to prescribe an abatement strategy that decreases ambient concentrations of pollutants by decreasing the emission rates. In contrast, emission rates from agricultural operations (including AFOs) are highly variable within a 24-hour period and over the course of a year.

Air quality management depends on accurate measurements of 1- to 24-hour average concentrations of some pollutants (e.g., the six criteria pollutants of the NAAQS) and allowable amounts of emissions per year or day for others (see Table 6-1 and Appendixes G and H). Measurements of concentrations of pollutants from sources with continuous and nearly constant emission rates (e.g., coal-fired power plants), whether based on ambient or source sampling, can easily be used to regulate the source. (It is assumed in simple rollback models that pollutant concentration is directly proportional to emission rates.) However, concentration measurements associated with sources with highly variable emission rates are a bigger problem. Ambient concentrations measured close to the source are affected by wind speed, direction, and atmospheric stability to vertical mixing. If the concentration measurement is to be used to calculate an emission factor, the differences between nonagricultural and agricultural emissions must be addressed in the protocol and methodologies used.

RECOMMENDATIONS:

- **Standardized protocols should be developed for both ambient and source concentration measurements for each substance of concern emitted from AFOs from both point and area sources.**
- **The accuracy and precision of analytical techniques for ammonia and odor have to be determined, including intercomparisons using controlled (i.e., synthetic air) and ambient air.**
- **Errors associated with measuring PM10 and PM2.5 concentrations from AFOs with federal reference method samplers must be corrected. Research priorities for PM (not strictly directed toward AFOs) have been previously suggested by the National Research Council (1998b).**
- **AFO air emission rates should be estimated using appropriate protocols for concentration measurements along with airflow rates in the case of barns or vertical turbulent diffusion characteristics in the case of area sources.**
- **This research program should allow for acquiring sufficient field data to statistically characterize the highly variable nature of emission rates from AFOs.**

NEED 2. The complexities of various kinds of air emissions and the temporal and spatial scales of their distribution make direct measurement at the individual farm level impractical other than in a research context. Research into the application of advanced three-dimensional modeling techniques accounting for transport over complex terrain under thermodynamically stable and unstable planetary boundary layer (PBL) conditions offers good possibilities for improving emission estimates from AFOs (Finding 6).

One of the ways in which emission factors can be estimated is through the use of dispersion modeling coupled with upwind and downwind ambient concentrations, to back-calculate the emission rates from the source. Gaussian-type dispersion models can be used if the terrain is relatively flat and the airflow is unimpeded; however, these ideal conditions rarely exist in actual settings. Where the terrain is complex, more sophisticated transport and dispersion modeling approaches should be used. Complex landscapes lead to complex solutions spaces, thus dictating the need for longer-term observations to characterize the transport-mixing and to identify its controls. Moreover, there is need for increased effort with multidimensional observations, such as with scanning lidar (light detection and ranging) that can characterize the evolving plume geometries. These data sets provide the basis for construction and testing of multiscale Eulerian modeling frameworks, where coarse Eulerian mesoscale models provide regional meteorological forcing and nested finer-scale Eulerian models predict plume characteristics over local surface features. There has however not been a widespread use of these advanced modeling techniques in the AFO regulatory context. The use of dispersion modeling with upwind and downwind concentration measurements also allows for emission rates calculated in this way to be reconciled with those that were estimated using other techniques (e.g., process-based models).

RECOMMENDATIONS:

- **Appropriate dispersion modeling procedures should be applied for back-calculating emission fluxes from point and area sources based on ambient sampling.**
- **Appropriate dispersion modeling procedures are needed for estimating downwind concentrations given accurate emission rates. If all AFOs are required to meet NAAQS and the only method for estimating downwind emission concentrations prior to construction is dispersion modeling, accurate estimates of downwind concentrations will be necessary. Existing ISCST3 (Industrial Source Complex Short Term) dispersion modeling is reported to give results that overestimate downwind concentrations by factors of 2.5 or greater (Meister et al., 2001; Fritz et al., 2001). This must be addressed in the research.**
- **Dispersion modeling would also permit sensitivity analyses of the impact of one concentrated source versus smaller sources distributed geographically across a region.**

NEED 3. Use of process-based modeling will help provide scientifically sound estimates of air emissions from AFOs for use in regulatory and management programs (Finding 9).

The committee recommends developing appropriate mathematical models to estimate emissions of several substances (e.g., nitrogen). The processes to develop such models and the laboratory and field studies necessary to parameterize them are discussed in Chapter 5. Specific research studies that are needed include the following.

Mathematical models and software applications should be developed to predict emissions from individual AFOs and multifarm animal production systems. These models and/or software applications could be used to evaluate individual AFOs and management technologies for their environmental impacts and to develop effective control strategies. Different models that may be developed should be compared to one another, and to empirical observations, to evaluate their accuracy and suitability for further research or application to control strategies.

In order to improve the accuracy of such models, studies are needed to determine the fractions of nitrogen lost as air emissions of NH_3, molecular nitrogen (N_2), N_2O, or NO from animal housing, anaerobic and aerobic manure storage, dry manure combustion, biogas generation, combustion of biogas, constructed or natural wetlands, and cropping systems or pasture. Until better studies are available, models will be required to use research, that may not accurately represent management options.

Better estimates are needed of the rates (amount or fraction per unit time) and fluxes (amount or fraction per unit time per unit surface area) of ammonia volatilization from animal housing, manure storage, and cropland, and rates of mineralization of organic nitrogen to ammonia. The importance of factors (e.g., temperature, pH) that affect these rates must be determined to improve mathematical models. This research should report the quantities and concentrations of various forms of nitrogen in storage or applied to land so that models can be based on mass balance kinetics.

Models to predict methane and hydrogen sulfide emissions from manure storage must be developed along with empirical research to parameterize them. For sulfide, the rates of emissions and the factors that affect those rates have to be determined for different ambient conditions.

While it is feasible to quantify the nitrogen, sulfur, and carbon excretion from animals on most AFOs, estimates of typical excretion rates are needed to make models that can be employed on a broad range of farms, for global inventories, or for screening of AFOs. Such estimates can be developed using existing feeding recommendations and surveys of feeding practices. For example, the American Society of Agricultural Engineers (ASAE) is currently revising some animal excretion and manure composition values (Powers and Koelsch, 2002).

The National Research Council (NRC) revises feeding recommendations for animal species of economic importance. It is critical to EPA and USDA efforts to ensure that these NRC reports are updated at regular intervals and that they include a review of all economically viable means to decrease the amounts of nitrogen, carbon, and sulfur in animal feces and urine. The reports should also include the predicted excretion of nitrogen, carbon, and sulfur in urine and feces and the emissions of methane directly from cattle.

RECOMMENDATIONS:

- **EPA and USDA should fund development of process-based mathematical models with mass balance constraints for nitrogen-containing compounds (i.e., NH_3, N_2O, NO), methane, and hydrogen sulfide to identify management changes that decrease emissions and to estimate the amounts by which these changes will decrease emissions to direct regulatory and management programs.**
- **EPA and USDA should investigate the potential use of a process-based model to estimate quantity of odor-causing compounds and potential management strategies to decrease their impacts.**
- **EPA and USDA should standardize procedures for estimating accurate emission inventories (annual emissions) from AFOs and multifarm animal production systems.**

NEED 4. Standardized methodology for odor measurement have not been adopted in the United States (Finding 5).

Mathematical models are needed to predict the emissions of odor-causing compounds such as ammonia, volatile organic compounds, and hydrogen sulfide. These models can be based on fundamental knowledge of the anaerobic and aerobic degradation processes. Research is needed on the contribution of different compounds to the strength and offensiveness of odors. Integration of this research would enable prediction of impacts of management changes on emissions of odor-causing compounds and the strength and types of odor. Once such models have been developed, they should be evaluated by comparing results to empirical observations to improve them and determine their suitability for further research as well as application in control strategies.

RECOMMENDATIONS:

- **Sampling and analysis techniques for odors and their individual constituents should be standardized.**
- **New techniques should be developed for establishing correlations between odor and individual compounds and combinations of microbial products.**
- **Studies of the microbial aspects of odor should be conducted. Ammonia, VOCs, and hydrogen sulfide are the major volatile components from livestock production facilities with appreciable odor. Zahn et al. (1997) established a significant correlation between air concentrations of VOCs and offensiveness of odors in swine production facilities. Therefore the formation and emission of VOCs may have a direct influence on the odors released. Studies to determine the diversity of the predominant microorganisms at AFOs might provide valuable information on how to control odors.**

NEED 5. Measurement protocols, control strategies and management techniques must be emission and scale specific (Finding 3).

Mathematical models and software applications should be developed to estimate the contribution of individual AFOs to total emissions from multifarm animal production systems. The supporting research necessary to develop these models is described in Need 3 above.

RECOMMENDATIONS:

- **Abatement and/or management strategies should be developed that will effectively decrease AFO emission rates. This research should include the technical description and operational details of the equipment and/or strategy with associated costs of implementation. The**

results should allow for cost-effective emission-limiting practices and/ or abatement strategies for regulatory agencies.

- **Control strategies aimed at decreasing unaccounted-for nitrogen from total animal production systems can be designed and implemented now. These strategies can include both performance standards based on individual farm calculations of nitrogen balance and technology standards to decrease total system nitrogen balance by quantifiable amounts.**
- **A systems approach, which integrates animal and crop production systems both on- and off-farm (e.g., imported feeds and exported manure), is necessary to evaluate air emissions from the total animal production system (Finding 8).**

LONG-TERM RESEARCH

The production of animal products (meat, milk, and eggs) in the United States has been increasing. Over the period 1990 to 2000, the production of beef, pork, and broilers increased by ~13 percent, 19 percent, and 67 percent, respectively. Production is projected to continue to increase through the next decade. Over the period 2000 to 2010, beef, pork, and broiler production are expected to increase by ~8 percent, 16 percent, and 32 percent, respectively (FAPRI, 2001), with much of the increase probably occurring in concentrated animal feeding operations (CAFOs). Given the likelihood of continued increases in the production of animal products in the United States and the potential impacts on both people and ecosystems, it is important to step back from the immediate need to measure AFO atmospheric emissions and look at longer time intervals in designing a research plan. A long-term research program would consist of a plan that is continuous and builds on the knowledge that is produced.

Defining AFOs

It is also necessary to take a broader approach in the definition of AFOs. EPA and USDA define an AFO as the location at which animals are fed and manure is stored and treated and the adjoining land to which the manure is applied. For the long-term research recommendations, the committee suggests expanding the definition to include remote locations where feed is grown and to which wastes are transferred, whether the wastes are stored or utilized by the AFO operator or by a contractor. The emissions from crop production, whether using animal manure or not, must also be considered. These crops may be consumed by animals or humans, with by-products consumed by animals. Thus, animal feed production may contribute to emissions or decrease them by using what would otherwise be considered a waste product. This expanded definition is necessary to ensure that we account for all emissions of the waste to the environment and provide a train of accountability for its proper management. The committee's definition thus has four components—inputs, recycling, waste outputs, and product outputs (Figure 7-1). Inputs include any materials used in support of animal production that enter the AFO system. In the case of feed, the input can be primary feed (e.g., corn produced for animal feed) or secondary feed (e.g., a waste product from a different type of agroecosystem). Water can contain sulfur as dissolved sulfate salts. Waste products are the nitrogen-, carbon-, sulfur-, and phosphorus-containing compounds in manure and mortalities. Recycling is the use of waste products at the AFO, at a different agriculture operation, or as a feedstock for another process. Products are meat, milk, eggs, live animals, and harvested crops.

In addition to proposing research for each component of the AFO system, the committee also recommends research on the whole system and on the impacts of emissions from AFOs on people and ecosystems. The overall goal is to decrease these emissions to such an extent that their impacts on people and ecosystems are minor relative to those from other sources.

Long-Term Research Recommendations

Inputs

Emissions of many pollutants from AFOs can be decreased if elemental (N, C, S, and P) inputs are decreased. If animals use nutrients that they consume more

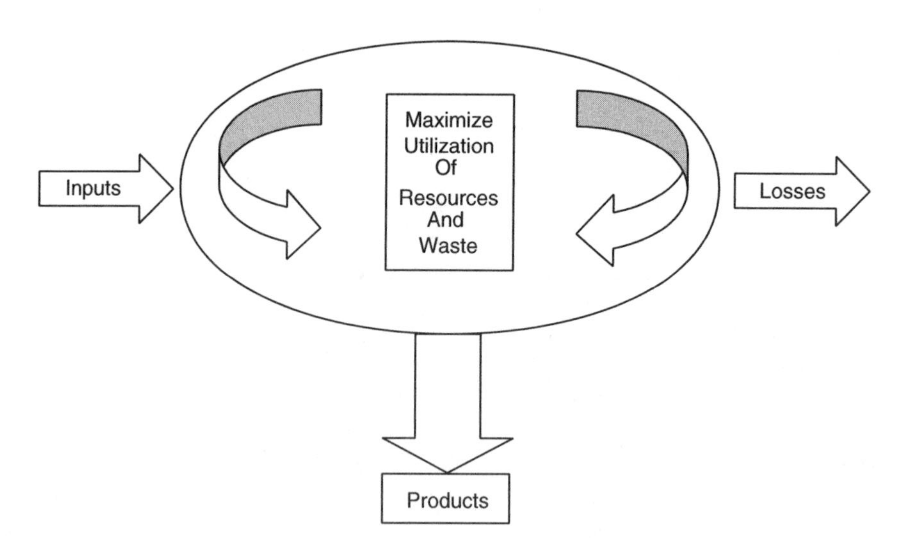

FIGURE 7-1 Animal feeding operations system (animals plus associated cropland).

efficiently, less will be excreted. For example, Kohn et al. (1997) showed for dairy farms that improvements in animal diet and management can increase the conversion of feed nitrogen to animal product by 50 percent, can increase total farm nitrogen efficiency by 48 percent, and can decrease nitrogen emissions per unit of product by 36-40 percent. The use of diet and management controls to increase total farm nutrient efficiency has received only limited attention. Improved nutrient utilization also decreases the need for the use of chemical fertilizer in feed crop production and the associated nutrient emissions to the environment. Several methods to do so have been reviewed (Klopfenstein et al., 2002). The approaches may be categorized as methods to (1) feed animals closer to their nutrient requirements, (2) increase production per animal, and (3) improve metabolism to increase efficiency of nutrient utilization. Feeding closer to animal requirements requires understanding what the animal's nutrient requirements actually are and how to best meet them with available forages, grains, and by-products. Since animal and crop production continually changes, these feeding requirements must be continually updated. The National Research Council Nutrient Requirement Series (e.g., 1994, 1998a, 2000, 2001a) plays a critical role in providing the foundation for EPA and USDA efforts in addressing AFO concerns by evaluating the current scientific knowledge and updating feeding recommendations for major species of domestic animals. Increasing production per animal (i.e., faster growth rates or higher milk or egg production) decreases the fixed costs of production per animal (primarily growing replacement stock and maintenance of basal activity) and thus improves the efficiency of nutrient utilization. There has been little progress in changing metabolism to increase production efficiency and decrease excretion. However, given current environmental concerns, such research may be warranted.

To increase the potential of this type of management to decrease emissions in general, and air emissions specifically, the committee recommends the following:

- Establish a coordinated research program based on the best available science and up-to-date estimates of animal nutrient requirements to determine how to optimize animal feeding so as to minimize waste.

The process of making and distributing nitrogen fertilizer uses a great deal of fossil fuel and results in carbon dioxide emissions to air. In addition, nitrogen fertilizers are taken up by crops at low efficiencies (i.e., 35-50 percent), with the remainder lost to the environment. Manure nitrogen is used at even lower efficiencies. Long-term research is needed on fertilizers that become available to crops when needed but otherwise remain stable in soils, to increase manure nitrogen availability and to increase the use of legume-type nitrogen fixation that can proceed with renewable energy with high efficiencies to produce fixed nitrogen.

Recycling

The use of products from manure is not new and crops have been manured for millennia. However, there are opportunities for reuse of manure beyond the usual. For example, Cowling et al. (2002), in an assessment for the State of North Carolina, recommended that swine waste be used to produce *energy* in the form of methane, biogas, diesel fuel, or electricity for direct on-farm purposes; *synthetic growth media* for high-value ornamental plants, or soil amendments for residential or commercial landscaping purposes; *nitrogen- and phosphorus-rich fertilizer materials* for direct application to crops such as corn, cotton, sweet potatoes, and so forth, or for fast-growing pine and/or hardwood plantations; *feed materials and nutritional supplements* to enhance feed conversion efficiency in fish, poultry, and livestock production (when recycled as animal feed, the potential spread of animal diseases must be considered); and *protein products for industrial applications* including industrial antibodies and enzymes used in detergents, recycling, and processing of pulp, paper, textile, and chemical products.

These ideas and others will require significant investments in research and development to move them to practical application. In that regard, the committee recommends the following:

- Establish a coordinated research program to determine how to convert waste from AFOs to usable products. The program should include the following:
  1. research into improved utilization of manure as fertilizer for growing crops;
  2. research into improved management of manure and other wastes derived from organic sources (municipal solid waste, sewage sludge, yard waste, wood, pulp, and paper by-products, food and food processing waste, etc.); and
  3. research on mechanical, chemical, and biological (microbial, aquatic, and terrestrial plants, etc.) methods of separating, concentrating, or otherwise refining and obtaining nutrients and energy from organic material. (An important method of decreasing reliance on natural ecosystems to process and recycle wastes from human activities is to contain these processes and recycle the nutrients and energy directly to productive uses. This recycling may decrease the need for producing more new reactive nitrogen and also decrease the need for mining and dispersing minerals and hydrocarbons.)

Waste Outputs

Because of the nature of AFOs, recycling of wastes will remain incomplete and there will be emissions to the environment. It is thus prudent to convert these wastes to innocuous forms through treatment. For nitrogen-containing species, this means increasing the potential for converting reactive nitrogen to N_2 at the AFO; for sulfur-containing species, it means converting reduced sulfur compounds to sulfate. The committee recommends the following:

- Establish a coordinated research program to determine how to convert unusable waste from AFOs to innocuous substances.
- There is a need for research on cost-effective management techniques for biological wastes that promote the return of reactive nitrogen to the atmosphere as harmless N_2 while decreasing the potential to form N_2O.
- Control strategies aimed at decreasing unaccounted-for nitrogen from total animal production systems can be designed and implemented now. These strategies can include both performance standards based on individual farm calculations of nitrogen balance and technology standards to decrease total system nitrogen balance by quantifiable amounts.
- Standard measurement protocols should be developed and used for quantifying the effects of alternative manure management practices and treatment technologies in decreasing emissions. Research should be conducted to quantify the effects of decreases in emissions.
- A critical area of research is innovative methods of waste handling. For example, if waste were stored in a closed system, not only would unplanned emissions to the environment not occur, but the material could more easily be converted into usable products (e.g., methane, compost) or transformed into an innocuous material (e.g., N_2).

Product Outputs

Given past successes, continued improvements in uptake efficiency of nitrogen, carbon, sulfur, and phosphorus by alterations of diets and feeding practices are likely. It is also likely that most of the material that enters the animal will be discharged as urine or feces; when combined with other wastes (e.g., bedding materials), the result is that AFOs will continue to be significant waste generators. The committee recommends the following:

- A coordinated research program needs to be established to determine how to use AFO products more efficiently to decrease loss of protein between the AFO and the consumer.
- Continued research is necessary to increase the efficiency of livestock production and to decrease emissions of pollutants to the environment through the following means:
  1. genetic selection of livestock for rapid growth: high rates of reproduction; good health and low mortality; high production of meat, milk,

and eggs; and highly efficient conversion of feed (high productivity implies less resource consumption per unit of product);

2. livestock nutrition: continued research into improved diet formulation and the use of new ingredients to increase the digestibility of feeds and the fraction of consumed nutrients going into products; work in this area will include evaluation of alternative processing methods; the use of enzymes, amino acids, ionophores, and so forth;
3. livestock health and welfare: continued research into improved health maintenance for livestock to decrease the incidence of morbidity and mortality (thereby decreasing the unproductive consumption of resources and decreasing the distribution of disease organisms); this research may also decrease the release of hormones and antibiotics into the general environment; and
4. livestock housing: continued research into improved housing design (indoor and outdoor) for livestock to improve their environment (clean, dry, at an appropriate temperature, adequately ventilated, low stress, low hazard, etc.) and minimize resource use (water, energy, feed, labor, land, capital, etc.) per unit of product.

Impacts

The emissions of nitrogen-, carbon-, sulfur-, and phophorus-containing substances from AFOs have impacts on people and ecosystems. Long-term research is needed on the following:

- A significant commitment of resources should be made to provide detailed scientific information on the contribution of AFO emissions to health and environmental effects (recommendation from Finding 1).
- Research on how to integrate regulatory and management programs to decrease air emissions with other environmental (e.g., water quality) and economic (e.g., affordable food production) programs should be developed (recommendation from Finding 8).
- Research is needed on the epidemiology and assessment of exposure to bioaerosols.
- The focus of this report is on substances that the committee was asked to evaluate. However, these are not the only chemical species that AFOs contribute to the environment and that have potential effects on humans and ecosystems. The committee proposes research to follow the fate of phosphorus and trace metals (e.g., zinc, copper) fed to animals and to address their long-term impacts on people and ecosystems and, in the case of trace metals, their potential for accumulation in agricultural soils.

System

The research recommended above is focused on individual components of the system (Figure 7-1) and the impacts of resultant emissions. Research is also needed on the system as a whole, including how to provide incentives to AFO operators and education to protein producers and consumers, specifically:

- USDA, EPA, and other relevant organizations (e.g., universities) should initiate and conduct coordinated research to determine which emissions from animal production systems (i.e., using systems analysis) are detrimental to the environment or public health and develop technologies to decrease their release. The overall research program should include research to optimize inputs to AFOs, optimize recycling of materials within AFOs, and significantly decrease releases to the environment (recommendation from Finding 11).
- In addition to process-level research on how to decrease emissions of deleterious material from AFOs, the committee also proposes a long-term educational program for protein producers and consumers. The focus of the program would be to educate both groups on the importance of AFOs and on the health and ecosystem consequences of inadvertent emissions of pollutants from them. From this balanced perspective, the protein producer would develop a broader view of the importance of controlling emissions. The protein consumer would develop an understanding of how much protein is required, how much protein the average person consumes, and the consequences of protein overconsumption.
- USDA and EPA should develop programs and markets that create positive incentives for producers to adopt and manage air emission mitigation and control. Research and development are required to develop efficient and effective programs (recommendation from Finding 13).
- The AFO system that is the focus of this report includes livestock housing and feeding, manure storage and treatment, and manure application to lands owned by the AFO operator. On the broader scale, there are other components that must be included when developing a long-term research program. Within the AFO, these include, but are not limited to, emissions from combustion at the AFO (e.g., diesel pumps). External to the AFO, they include agroecosystems that provide the animal feed (e.g., grain and food by-products) and off-site agroecosystems that use the manure disposed by the AFO. The committee proposes research to examine AFOs as defined in this broader context in order to develop a true systems approach to AFO management.

Concentration and Emission Measurements

Research required to obtain scientifically sound estimates of air emissions from AFOs is addressed primarily in the previous section on short-term research recommendations. Findings of the short-term research will determine the course of long-term research in this area.

SUMMARY

The short- and long-term research recommendations address issues that are associated with regulation of AFO emissions based on priorities established in Finding 2. The committee recommends that a process-based modeling approach be used as a replacement for the emission factor approach to limit and/or regulate for most AFO emissions. The critical short-term research needs include concentration measurements, dispersion modeling (local and air shed), odor measurement and characterization, and abatement and/or management strategies. The critical long-term research needs are focused on an integrated program to reduce the losses of materials from AFOs to the environment by more efficient use of input materials and increased recycling of nitrogen-, carbon-, sulfur-, and phosphorus-containing materials within the AFOs. Wastes that cannot be used within the system should either be converted into a product that can be used by another sector of society or be converted into a material that will not harm people or the environment (e.g., N_2). The goal of these recommendations is to address the lack of technical data, procedures, equipment, management practices, and abatement equipment needed to limit AFO emissions while maintaining a viable AFO industry. The implementation of these recommendations requires willingness within EPA and USDA not to do business as usual, with the same programs and same resource allocation. Rather, new partnerships between the agencies and with other groups need to be forged and significant increases in funding must be allocated.

8

Conclusions

Estimating air emission rates and concentrations from animal feeding operations (AFOs) to guide regulatory and management programs is complicated. The large number and wide variety of operations, even among those for a single livestock type, limit the usefulness of averages in attributing emissions to specific farms. Differences among farms in management practices, such as manure handling, topography, and climate, add to the complexities. Differences in meteorological conditions over time and space also pose difficulties in estimating air emissions.

The U.S. Environmental Protection Agency's (EPA's) interest has usually focused on estimates of atmospheric concentrations at the "fence line" of farms with large animal feeding operations. This has led to complications because such concentrations are determined both by the generation of emissions and by their dispersion, which is affected by meteorological conditions. Measuring emissions during constant meteorological conditions as a basis for their estimation at other times and locations is difficult.

The committee has addressed this problem and concluded that both the EPA and the U.S. Department of Agriculture (USDA) would be better served by an approach to estimating these emissions that is quite different from that used to date. This approach is termed "process-based" modeling, which the committee expects will largely replace the current "emission factor" approach. The committee's findings and major recommendations that provide the rationale for the approach are listed below.

Recommendations for a short-term research agenda (four to five years) are intended to provide EPA and USDA with scientifically sound information for their decisions within that time frame. For the longer term (20-30 years), a research program is proposed that will ultimately lead to greatly decreased emissions to the environment of the constituents of air emissions as well as other losses from AFOs while maintaining a high level of production.

Effective and early adoption of the committee's recommendations will require the commitment of EPA and USDA to a course of action that will result in scientifically credible information for their air quality programs. It will require a willingness to change the direction of some current programs and to support a vigorous, expanded research agenda. The committee believes that the results will be of great value, not only in protecting health and the environment, but in reducing costs, and improving the health and welfare of the animals.

SETTING PRIORITIES

Air emissions from animal feeding operations are of varying concern at different spatial scales, as shown in Table 8-1 **(Finding 2, pp. 4, 71)**.

RECOMMENDATION: These differing effects, concentrations, and spatial distributions lead to a logical plan of action for establishing research priorities to provide detailed scientific information on the contri-

TABLE 8-1 Committee's Scientific Evaluation of the Potential Importance of AFO Emissions at Different Spatial Scales

Emissions	Global, National, Regional	Local, Property Line, Nearest Dwelling	Primary Effects of Concern
NH_3	Major[a]	Minor	Atmospheric deposition, haze
N_2O	Significant	Insignificant	Global climate change
NO_x	Significant	Minor	Haze, atmospheric deposition, smog
CH_4	Significant	Insignificant	Global climate change
VOCs[b]	Insignificant	Minor	Quality of human life
H_2S	Insignificant	Significant	Quality of human life
PM10[c]	Insignificant	Significant	Haze
PM2.5[c]	Insignificant	Significant	Health, haze
Odor	Insignificant	Major	Quality of human life

[a]Relative importance of emissions from AFOs at spatial scales based on committee's informed judgment on known or potential impacts from AFOs. Rank order from high to low importance is major, significant, minor, insignificant. While AFOs may not play an important role for some of these, emissions from other sources alone or in aggregate may have different rankings. For example VOCs and NO_x play important roles in the formation of tropospheric ozone; however the role of AFOs is likely insignificant compared to other sources.

[b]Volatile organic compounds.

[c]Particulate matter. PM10 and PM2.5 include particles with aerodynamic equivalent diameters up to 10 and 2.5 mm, respectively.

butions of AFO emissions to potential effects and the subsequent implementation of control measures. USDA and EPA should first focus their efforts on measurement and control of those emissions of major concern.

Measurement protocols, control strategies and management techniques must be emission and scale specific **(Finding 3, pp. 5, 71).**

RECOMMENDATIONS:

- **For air emissions important on a global or national scale (i.e., ammonia [NH_3] and the greenhouse gases, methane [CH_4], and nitrous oxide [N_2O]), the aim is to control emissions per unit of production (kilogram of food produced) rather than emissions per farm. Where the environmental and health benefits outweigh the costs of mitigation it is important to decrease the aggregate emissions. In some geographic regions, aggregate emission goals may limit the number of animals produced in those regions.**
- **For air emissions important on a local scale (hydrogen sulfide [H_2S], particulate matter [PM], and odor), the aim is to control ambient concentrations at the farm boundary and/or nearest occupied dwelling. Standards applicable to the farm boundary and/or nearest occupied dwelling must be developed.**
- **Monitoring should be conducted to measure concentrations of air pollutants with possible health concerns at times when they are likely to be highest and in places where the densities of animals and humans, and typical meteorological conditions, are likely to result in the highest degree of human exposure.**

ESTIMATING AIR EMISSIONS

There is a general paucity of credible scientific information on the effects of mitigation technology on concentrations, rates, and fates of air emissions from AFOs. However, the implementation of technically and economically feasible management practices (e.g., manure incorporation into soil) designed to decrease emissions should not be delayed **(Finding 4, pp. 6, 72).**

RECOMMENDATION: Best management practices (BMPs) aimed at mitigating AFO air emissions should continue to be improved and applied as new information is developed on the character, amount, and dispersion of these air emissions, and on their health and environmental effects. A systems analysis should include impacts of a BMP on other parts of the entire system.

Standardized methodology for odor measurement have not been adopted in the United States **(Finding 5, pp. 7, 86)**.

RECOMMENDATIONS:

- **Standardized methodology should be developed in the United States for objective measurement techniques to correspond to subjective human response.**
- **A standardized unit of measurement of odor concentration should be adopted in the United States.**

Scientifically sound and practical protocols for measuring air concentrations, emission rates, and fates are needed for the various elements (nitrogen, carbon, sulfur), compounds (e.g., NH_3, CH_4, hydrogen sulfide [H_2S]), and particulate matter **(Finding 7, pp. 8, 96)**.

RECOMMENDATIONS:

- **Reliable and accurate calibration standards should be developed, particularly for ammonia.**
- **Standardized sampling and compositional analysis techniques should be provided for PM, odor, and their individual components.**
- **The accuracy and precision of analytical techniques for ammonia and odor should be determined, including intercomparisons on controlled (i.e., synthetic) and ambient air.**

SYSTEMS APPROACH

Much confusion exists about the use of the term "animal unit" because EPA and USDA define animal unit differently **(Finding 1, pp. 4, 35)**.

RECOMMENDATION: Both EPA and USDA should agree to define animal in terms of animal live weight rather than any arbitrary definition of animal unit.

Estimating air emissions from AFOs by multiplying the number of animal units by existing emission factors is not appropriate for most substances **(Finding 8, pp. 9, 101)**.

RECOMMENDATION: The science for estimating air emissions from individual AFOs should be strengthened to provide a broadly recognized and acceptable basis for regulations and management programs aimed at mitigating the effects of air emissions.

Use of process-based modeling will help provide scientifically sound estimates of air emissions from AFOs for use in regulatory and management programs **(Finding 9, pp. 9, 103)**.

RECOMMENDATIONS:

- **EPA and USDA should use process-based mathematical models with mass balance constraints for nitrogen-containing compounds, methane, and hydrogen sulfide to identify, estimate, and guide management changes that decrease emissions for regulatory and management programs.**
- **EPA and USDA should investigate the potential use of a process-based model to estimate mass emissions of odorous compounds and potential management strategies to decrease their impacts.**
- **EPA and USDA should commit resources and adapt current or adopt new programs to fill identified gaps in research to improve mathematical process-based models to increase the accuracy and simplicity of measuring and predicting emissions from AFOs (see short-term and long-term research recommendations).**

A systems approach, which integrates animal and crop production systems both on and off (imported feeds and exported manure) the AFO, is necessary to evaluate air emissions from the total animal production system **(Finding 10, pp. 10, 114)**.

RECOMMENDATION: Regulatory and management programs to decrease air emissions should be integrated with other environmental (e.g., water quality) and economic considerations to optimize public benefits.

RESEARCH NEEDS

The complexities of various kinds of air emissions and the temporal and spatial scales of their distribution make direct measurement at the individual farm level impractical other than in a research setting. Research into the application of advanced three-dimensional modeling techniques accounting for transport over complex terrain under thermodynamically stable and unstable planetary boundary layer (PBL) conditions offers good possibilities for improving emissions estimates from AFOs **(Finding 6, pp. 7, 95)**.

RECOMMENDATION: EPA should develop and carry out one or more intensive field campaigns to evaluate the extent to which ambient atmospheric concentrations of the various species of interest are consistent with estimated emissions and to understand how transport and

chemical dynamics shape the local and regional distribution of these species.

Nitrogen emissions from AFOs and total animal production systems are substantial and can be quantified and documented on an annual basis. Measurements and estimates of individual nitrogen species components (i.e., NH_3, molecular nitrogen [N_2], N_2O, and nitric oxide [NO]) should be made in the context of total nitrogen losses **(Finding 11, pp. 10, 115)**.

RECOMMENDATION: Control strategies aimed at decreasing emissions of reactive nitrogen compounds (Nr) from total animal production systems should be designed and implemented now. These strategies can include both performance standards based on individual farm calculations of nitrogen balance and technology standards to decrease total system emissions of reactive nitrogen compounds by quantifiable amounts.

USDA and EPA have not devoted the necessary financial or technical resources to estimate air emissions from AFOs and develop mitigation technologies. The scientific knowledge needed to guide regulatory and management actions requires close cooperation between the major federal agencies (EPA, USDA), the states, industry and environmental interests, and the research community, including universities **(Finding 12, pp. 11, 153)**.

RECOMMENDATIONS:

- **EPA and USDA should cooperate in forming a continuing research coordinating council (1) to develop a national research agenda on issues related to air emissions from AFOs in the context of animal production systems and (2) to provide continuing oversight on the implementation of this agenda. This council should include representatives of EPA and USDA, the research community, and other relevant interests. It should have authority to advise on research priorities and funding.**
- **Exchanges of personnel among the relevant agencies should be promoted to encourage efficient use of personnel, broadened understanding of the issues, and enhanced cooperation among the agencies.**
- **For the short term, USDA and EPA should initiate and conduct a coordinated research program designed to produce a scientifically sound basis for measuring and estimating air emissions from AFOs on local, regional, and national scales.**
- **For the long term, USDA, EPA, and other relevant organizations should conduct coordinated research to determine which emissions (to water and air) from animal production systems are most harmful to the environment and human health and to develop technologies to decrease their releases into the environment. The overall research**

program should include research to optimize inputs to AFOs, optimize recycling of materials, and significantly decrease releases to the environment.

Setting priorities for both short- and long-term research on estimating air emission rates, concentrations, and dispersion requires weighing the potential severity of adverse impacts, the extent of current scientific knowledge about them, the potential for advancing scientific knowledge, and the potential for developing successful mitigation and control strategies **(Finding 13, pp. 12, 154)**.

RECOMMENDATIONS:

- **Short-term research priorities should improve estimates of emissions on individual AFOs including effects of different control technologies:**
  1. **Priority research for emissions important on a local scale should be conducted on odor, PM, and H_2S (also see Finding 2).**
  2. **Priority research for emissions important on regional, national, and global scales should be conducted on ammonia, N_2O, and methane (also see Finding 2).**
- **Long-term research priorities should improve understanding of animal production systems and lead to development of new control technologies.**

References

Abbasi, M.K., Z. Shah, and W.A. Adams. 1997. Concurrent nitrification and denitrification in compacted grassland soil. Pp. 47-54 in Gaseous Nitrogen Emissions from Grasslands, S.C. Jarvis and B.F. Pain (eds.). New York: CAB International.

Alexander, M. 1977. Introduction to Soil Microbiology, 2nd ed. New York: John Wiley & Sons.

Amon, M., M. Dobeic, R.W. Sneath, V.R. Phillips, T.H. Misselbrook, and B.F. Pain. 1997. A farm-scale study on the use of clinoptilolite zeolite and Deodorase for reducing odour and ammonia emissions from broiler houses. Bioresource Technology 61:229-237.

Amon, B., T. Amon, J. Boxberger, and C. Alt. 2001. Emissions of NH_3, N_2O and CH_4 from dairy cows housed in a farmyard manure tying stall. Nutrient Cycling in Agroecosystems 60:103-113.

Andersson, M. 1998. Reducing ammonia emissions by cooling of manure in manure culverts. Nutrient Cycling in Agroecosystems 51:73-79.

Aneja, V.P., J.P. Chauhan, and J.T. Walker. 2000. Characterization of atmospheric ammonia emissions from swine waste storage and treatment lagoons. Journal of Geophysical Research 105:11535-11545.

Aneja, V.P., B. Bunton, J.T. Walker, and B.P. Malik. 2001. Measurement and analysis of atmospheric ammonia emissions from anaerobic lagoons. Atmospheric Environment 35:1949-1958.

Angel, R. 2000. Feeding Poultry to Minimize Manure Phosphorus. Pp. 185-195 in Proceedings from Managing Nutrients and Pathogens from Animal Agriculture. Natural Resource, Agriculture, and Engineering Service (NRAES-130). Camp Hill, Penn.

Apel, E.C., J.G. Calvert, and F.C. Fehsenfeld. 1994. The Nonmethane Hydrocarbon Intercomparison Experiment (NOMHICE): Tasks 1 and 2. Journal of Geophysical Research 99:16651-16664.

Apel, E.C., J.G. Calvert, and T.M. Gilpin. 1999. The Nonmethane Hydrocarbon Intercomparison Experiment (NOMHICE): Task 3. Journal of Geophysical Research 104:26069-26086.

ARC (Agricultural Research Council). 1981. Pp. 307 in The Nutrient Requirements of Pigs: Technical Review, 2nd ed. Farnham Royal, U.K.: Commonwealth Agricultural Bureaux.

Arogo, J., R.H. Zhang, G.L. Riskowski, and D.L. Day. 1999. Mass transfer coefficient for hydrogen sulfide emission from aqueous solutions and liquid swine manure. Transactions of the American Society of Agricultural Engineers 42:1455-1462.

Arogo, J., R.H. Zhang, G.L. Riskowski, and D.L. Day. 2000. Hydrogen sulfide production from stored liquid swine manure: A laboratory study. Transactions of the American Society of Agricultural Engineers 43:1241-1245.

Arogo, J., P.W. Westerman, A.J. Heber, W.P. Robarge, and J.J. Classen. 2001. Ammonia Emissions from Animal Feeding Operations. Ames, Iowa: National Center for Manure and Animal Waste Management and Midwest Plan Services.

ARS (Agricultural Research Service). 2002. ARS National Programs. Available on-line at http://www.nps.ars.usda.gov/ [November 2002].

ASAE (American Society of Agricultural Engineers). 1971. Livestock waste management and pollution abatement. In Proceedings of the Second International Symposium on Livestock Wastes, St. Joseph, Mich. 360 pp.

ASAE (American Society of Agricultural Engineers). 1999. Control of manure odors. ASAE EP-379.2. Pp. 655-657 in ASAE Standards - 1999, 46th ed., St. Joseph, Mich.

ATSDR (Agency for Toxic Substances and Disease Registry). 1990. Toxicological Profile of Ammonia. NTIS Order number PB/91/180315/AS. Atlanta, Ga.: Centers for Disease Control and Prevention.

ATSDR (Agency for Toxic Substances and Disease Registry). 1999. Toxicological Profile of Hydrogen Sulfide. NTIS Order number PB/99/166696. Atlanta, Ga.: Centers for Disease Control and Prevention.

Auvermann, B., R. Bottcher, A. Heber, D. Meyer, C.B. Parnell, Jr., B. Shaw, and J. Worley. 2001. Particulate Matter Emissions from Confined Animal Feeding Operations: Management and Control Measures. Ames, Iowa: National Center for Manure and Animal Waste Management and Midwest Plan Services.

Avery, G.L., G.E. Merva, and J.B. Garrish. 1975. Hydrogen sulfide production in swine confinement units. Transactions of the American Society of Agricultural Engineers 18:149-151.

AWWA (American Water Works Association). 1990. Water Quality and Treatment. New York: McGraw-Hill.

Banwart, W.L., and J.M. Bremner. 1975. Identification of sulfur gases evolved from animal manures. Journal of Environmental Quality 4:363-366.

Barnebey-Cheney. 1987. Scentometer: An Instrument for Field Odor Measurement. Bulletin T-748 Barnebey-Cheney Activated Carbon and Air Purification Equipment Co., Columbus, Ohio. 3 pp.

Barth, C.L., and L.B. Polkowski. 1974. Identifying odorous components of stored dairy manure. Transactions of the American Society of Agricultural Engineers 17:737-741.

Barth, C.L., L.F. Elliot, and S.W. Melvin. 1984. Using Odor Control Technology to Support Animal Agriculture. Transactions of the American Society of Agricultural Engineers 27:859-864.

Battye, R., W. Battye, C. Overcash, and S. Fudge. 1994. Pp. 2-1 to 2-20 in Development and Selection of Ammonia Emission Factors. Durham, N.C.: EC/R Inc.

Beard, W.E., and W.D. Guenzi. 1983. Volatile sulfur compound from a redox-controlled cattle manure slurry. Journal of Environmental Quality 12:113-116.

Berges, M.G.M., and P.J. Crutzen. 1996. Estimates of global N_2O emissions from cattle, pig, and chicken manure, including a discussion of CH_4 emissions. Journal of Atmospheric Chemistry 24:241-269.

Bicudo, J.R., C.L. Tengman, D.R. Schmidt, and L.D. Jacobson. 2002. Ambient H_2S concentrations near swine barns and manure storages. Paper presented at 2002 ASAE Annual International Meeting, CIGR XVth World Congress, Chicago, Ill. July 28-31. American Society of Agricultural Engineers Paper No: 024059.

Billesbach, D.P., J.K. Kim, R.J. Clement, S.B. Verma, and F.G. Ullman. 1998. An intercomparison of two tunable diode laser spectrometers used for eddy correlation measurements of methane flux in a prairie wetland. Journal of Atmospheric and Oceanic Technology 15:197-206.

Blaxter, K.L., and J.L. Clappperton. 1965. Prediction of the amount of methane produced by ruminants. British Journal of Nutrition 19:511.

Blayney, D. 2002. The Changing Landscape of U.S. Milk Production. ERS Statistical Bulletin 978. Washington D.C.: Economic Research Service, United States Department of Agriculture.

Blayney, D., and A. Manchester. 2001. Milk Pricing in the United States. ERS AIB 761. Washington D.C.: Economic Research Service, United States Department of Agriculture.

Bouwman, A.F. 1996. Direct emission of nitrous oxide from agricultural soils. Nutrient Cycling in Agroecosystems 46:53-70.

Brady, L.J., D.R. Romsos, P.S. Brady, W.G. Bergen, and G.A. Leveille. 1978. The effects of fasting on body composition, glucose turnover, enzymes and metabolites in the chicken. Journal of Nutrition 108:648-657.

Brock, T.D., and M.T. Madigan. 1988. Biology of Microorganisms, 5th ed. Englewood Cliffs, NJ: Prentice Hall.

Brownell, F.W. 2001. The Clean Air Act. Pp. 193-245, in Environmental Law Handbook, 16th ed. Rockville, Md.: Government Institutes.

Burton, C.H. (ed.). 1997. Manure Management: Treatment Strategies for Sustainable Agriculture. Silsoe, Bedford, U.K.: Silsoe Research Institute.

Buser, M., C.B. Parnell, B.W. Shaw, and R. Lacey. 2002. PM10 sampler errors due to the interaction of particle size and sampler performance characteristics. Proceedings of the 2002 Beltwide Cotton Production Conferences. Memphis, Tenn.: National Cotton Council.

California Environmental Protection Agency. 1998. Environmental Monitoring Branch of the Department of Pesticide Regulation in Sacramento. Available on-line at http://www.cdpr.ca.gov/docs/pur/vocproj/voc_em.htm [March 2002].

California Environmental Protection Agency. 1999. Environmental Monitoring Branch of the Department of Pesticide Regulation in Sacramento. Available on-line at http://www.cdpr.ca.gov/docs/pur/vocproj/voc_em.htm [March 2002].

California Environmental Protection Agency. 2000. Environmental Monitoring Branch of the Department of Pesticide Regulation in Sacramento. Available on-line at http://www.cdpr.ca.gov/docs/pur/vocproj/voc_em.htm [March 2002].

Carlton, D.W., and J.M. Perloff. 1989. Modern Industrial Organization, 2nd ed. Glenview, Ill.: Scott Foresman. 852 pp.

Chang, C., C.M. Cho, and H.H. Janzen. 1998. Nitrous oxide emission from long-term manured soils. Soil Science Society of America Journal 62:677-682.

Chen, A., P.H. Liao, and K.V. Lo. 1994. Headspace analysis of malodorous compounds from swine wastewater under aerobic treatment. Bioresource Technology 49:83-87.

Chen, T.H., D.L. Day, and M.P. Steinberg. 1988. Methane production from fresh vs. dry dairy manure. Biological Wastes 24:297-306.

Childers, J.W., E. Thompson, D. Harris, D. Kirchgessner, M. Clayton, D. Natschke, and W. Phillips. 2001. Multi-pollutant concentration measurements around a concentrated swine production facility using open-path FTIR spectrometry. Atmospheric Environment 35:1923-1936.

Chow, J.C., J.G. Watson, Z. Lu, D.H. Lowenthal, C.A. Frazier, P.A. Solomon, R.H. Thuillier, and K. Magliano. 1996. Descriptive analysis of PM2.5 and PM10 at regionally representative locations during SJVAQS/AUSPEX. Atmospheric Environment 30:2079-2111.

Civerolo, K.L., and R.R. Dickerson. 1998. Nitric oxide soil emissions from tilled and untilled cornfields. Agricutlural and Forest Meteorology 90:307-311.

Clayton H., J. Arah, and K.A. Smith. 1994. Measurement of nitrous-oxide emissions from fertilized grassland using closed chambers. Journal of Geophysical Research-Atmospheres 99:6599-16607.

Colorado Air Quality Control Commission. 1999. Regulation No. 2: Odor Emission. Available online at http://www.cdphe.state.co.us/op/regs/airregs/100104aqccodoremission.pdf.

Cowling, E.B., J.N. Galloway, C.S. Furiness, M.C. Barber, T. Bresser, K. Cassman, J.W. Erisman, R. Haeuber, R.I. Howarth, J. Melillo, W. Moomaw, A. Mosier, K. Sanders, S. Seitzinger, S. Smeulders, R. Socolow, D. Walters, F. West, and Z. Zhu. 2002. Optimizing nitrogen management in food and energy production and environmental protection: Summary statement from the Second International Nitrogen Conference. In Optimizing Nitrogen Management in Food and

Energy Production and Environmental Protection. Contributed papers from the Second International Nitrogen Conference, Potomac, Md., October 14-18, 2001, J.N. Galloway, E.B Cowling., J.W Erisman., J. Wisniewski, and C. Jordan (eds.). Lisse/Abingdon/Exton, Pa./Tokyo: A.A. Balkema Publishers.

Cox, V. 1993. A Cost Analysis of Irrigating Swine Waste Effluent. Unpublished master's thesis. Department of Agricultural and Resource Economics, North Carolina State University, Raleigh.

Crites, R., and G. Tchobanogolous. 1998. Small and Decentralized Wastewater Management Systems. Boston: WCB McGraw Hill.

Crosley, D.R. 1996. The NO_y Blue Ribbon Panel. Journal of Geophysical Research 101:2049-2052.

CSREES (Cooperative State Research, Education, and Extension Service). 2001. Animal Waste Management, Available on-line at http://www.reeusda.gov/nre/water/ANWASTE/.

CSREES (Cooperative State Research, Education, and Extension Service). 2002. Impacts of the Research, Extension, and Education Activities of the Land-Grant University System. Available on-line at http://www.reeusda.gov/.

Davidson, E.A., and W. Klingerlee. 1997. A global inventory of nitric oxide emissions from soils. Nutrient Cycling Agroecosystems 48:37-50.

Dendooven, L., L. Duchateau, and J.M. Anderson. 1996. Gaseous products of the denitrification process as affected by the antecedent water regime of the soil. Soil Biology and Biochemistry 28:239-245.

Denmead, O.T. 1997. Progress and challenges in measuring and modelling gaseous nitrogen emissions from grasslands: An overview. Pp. 423-438 in Gaseous Nitrogen Emissions from Grasslands, S.C. Jarvis and B.F. Pain (eds.). New York: CAB International.

Dentener, F.J., and P.J. Crutzen. 1994. A 3-dimensional model of the global ammonia cycle. Journal of Atmospheric Chemistry 19:331-369.

Dewulf, J., and H. Van Langenhove. 1997. Analytical aspects for the determination and measurement data of 7 chlorinated C_1- and C_2-hydrocarbons and 6 monocyclic aromatic hydrocarbons in remote air masses: An overview. Atmospheric Environment 31:3291-3307.

Dinn, N.E., J.A. Shelford, and L.J. Fisher. 1998. Use of the Cornell Net Carbohydrate and Protein System and rumen-protected lysine and methionine to reduce nitrogen excretion from lactating dairy cows. Journal of Dairy Science 81:229-237.

DOE (U.S. Department of Energy). 2000. Emission of Greenhouse Gases in the United States 2000. Department of Energy Report: DOE/EIA-0573. Washington D.C.

Dou, Z., R.A. Kohn, J.D. Ferguson, R.C. Boston, and J.D. Newbold. 1996. Managing nitrogen on dairy farms: An integrated approach. 1. Model description. Journal of Dairy Science 79:2071-2080.

Dou, Z., J.D. Ferguson, L.E. Lanyon, and R.A. Kohn. 1998. Managing nitrogen on dairy farms: An integrated approach. 2. Field application. Agronomy Journal 90: 573-581

Drabenstott, M., M. Henry, and K. Mitchell. 1999. Where have all the packing plants gone? The new meat geography in rural America. Pp. 65-82 in Economic Review, third quarter, Kansas City: Federal Reserve.

Drynan, R., J.C. Williamson, P.W. Westerman, and S.R. Crane. 1981. Economic and Technical Evaluations of Swine Manure Management Systems Which Use Land Applications. Report No. 173. University of North Carolina: Water Resources Research Institute. Chapel Hill, N.C.

Dunlap, T.F., R.A. Kohn, G.E. Dahl, and M. Varner. 2000. The impact of somatotropin, milking frequency, and photoperiod on dairy farm nutrient flows. Journal of Dairy Science 83:968-976.

Eaton, D. L. 1996. Swine Waste Odor Compounds. Pioneer Hi-Bred International, Livestock Environmental Systems, Des Moines, Iowa. 14 pp.

Elliott, L.F., J.W. Doran, and T.A. Travis. 1978. A review of analytical methods for detecting and measuring malodors from animal wastes. Transaction of the American Society of Agricultural Engineers 21:130-135.

Ellis, S., J. Webb, T. Misselbrook, and D. Chadwick. 2001. Emissions of ammonia (NH_3), nitrous oxide (N_2O) and methane (CH_4) from a dairy hardstanding in the UK. Nutrient Cycling in Agrosystems 60:115-122.

ENS (Environmental News Service), 2002. Tyson Foods Charged with Superfund Violations. Available on-line at http://ens-news.com/ens/Feb2002/2002L-02-06-09.html.

EPA (U.S. Environmental Protection Agency). 1992. Global Methane Emissions from Livestock and Poultry Manure. EPA/4001/1-92/048. Washington, D.C.

EPA (U.S. Environmental Protection Agency). 1993a. Anthropogenic Methane Emissions in the United States: Estimates for 1990 Report to Congress. EPA 430-R-93-003. Atmospheric Pollution Prevention Division, EPA Office of Air and Radiation, Washington, D.C. Available on-line at http://www.epa.gov/ghginfo/reports.htm.

EPA (U.S. Environmental Protection Agency). 1993b. Guidance Specifying Management Measures for Sources of Nonpoint Pollution in Coastal Waters. EPA-840-B-93-001c. Available on-line at http://www.epa.gov?OWOW/NPS/MMGI/.

EPA (U.S. Environmental Protection Agency). 1995a. National Air Quality and Emissions Trends Report. Publication Number 454-R-96-005. Washington, D.C.

EPA (U.S. Environmental Protection Agency). 1995b. Compilation of Air Pollutant Emission Factors AP-42, 5th ed., Volume I: Stationary Point and Area Sources. Research Triangle Park, N.C.

EPA. (U.S. Environmental Protection Agency). 1998a. Solid Waste and Emergency Response EPA, RCRA, Superfund & EPCRA Hotline Training Module. EPA 540-R-98-022. Available on-line at http://www.epa.gov/superfund/contacts/sfhotlne/cerep.pdf [October 2002].

EPA (U.S. Environmental Protection Agency). 1998b. Draft Strategy for Addressing Environmental and Public Health Impacts from Animal Feeding Operations. Available on-line at http://www.epa.gov/npdes/pubs/astrat.pdf.

EPA (U.S. Environmental Protection Agency). 1998c. Compliance Assurance Implementation Plan for Animal Feeding Operations. Available on-line at http://es.epa.gov/oeca/strategy.html.

EPA (U.S. Environmental Protection Agency). 2000a. Inventory of U.S. Greenhouse Gas Emissions and Sinks (1990-1998). EPA 236-R-00-001. Washington, D.C.

EPA (U.S. Environmental Protection Agency). 2000b. Toxic Release Inventory Factsheet. Available on-line at http://www.epa.gov/tri/tri_program_fact_sheet.htm [November 2002].

EPA (U.S. Environmental Protection Agency). 2000c. Outreach Document for the U.S. EPA's Proposed Regulatory Changes to the 1) National Pollutant Discharge Elimination System Concentrated Animal Feeding Operation (CAFO) Regulations and 2) Effluent Limitation Guidelines for Feedlots. Washington, D.C.

EPA (U.S. Environmental Protection Agency). 2001a. Emissions from Animal Feeding Operations (Draft). EPA Contract No. 68-D6-0011. Washington, D.C. Available on-line at http://epa.gov/ttn/chief/ap42/ch09/draft/draftanimalfeed.pdf.

EPA (U.S. Environmental Protection Agency). 2001b. Proposed Regulations to Address Water Pollution from Concentrated Animal Feeding Operations. EPA 833-F-00-016. Washington, D.C.

EPA (U.S. Environmental Protection Agency). 2001c. Public Commenter's Guide to the Proposed New CAFO Regulations, Available on-line at http://www.epa.gov/npdes/pubs/cafocomguide.pdf.

EPA (U.S. Environmental Protection Agency). 2001d. State Compendium: Programs and Regulatory Activities Related to Animal Feeding Operations, Available on-line at http://cfpub.epa.gov/npdes/images/StateCompPDF/statecom.pdf [November 2002].

EPA (U.S. Environmental Protection Agency). 2002. Section 112 Hazardous Air Pollutants. Available on-line at http://www.epa.gov/ttn/atw/188polls.html [October 2002].

EPA (U.S. Environmental Protection Agency) and USDA (U.S. Department of Agriculture). 1998. Clean Water Action Plan. Available on-line at http://www.cleanwater.gov/action/.

Esler, M.B., D.W.T. Griffith, S.R. Wilson, and L.P. Steele. 2000. Precision trace gas analysis by FT-IR spectroscopy 1. Simultaneous analysis of CO_2, CH_4, N_2O and CO in air. Analytical Chemistry 72(1): 206-215.

European Committee for Standardization. 2001. European Standards. Available on-line at http://www.cenorm.be/catweb/ [November 2002].

European Committee for Standardization. 2002. EN 13725. Determination of odour concentration by dynamic olfactometry. Brussels: CEN Central Secretariat.

FAPRI (Food and Agriculture Policy Research Institute). 2001. World and U.S. Agricultural Outlook. Ames, Iowa. Available on-line at http://www.fapri.org/Outlook2001/outlook2001.htm [October 2002].

Farwell, S.O., A.E. Sherrard, and M.R. Pack. 1979. Sulfur compounds volatilized from soils at different moisture contents. Soil Biology and Biochemistry 11:411-415.

Fehsenfeld, F.C., R.R. Dickerson, G. Huebler, W.T. Luke, L.J. Nunnermacker, E.J. Williams, J. Roberts, J.G. Calvert, C. Curran, A.C. Delany, C.S. Eubank, D.W. Fahey, A. Fried, B. Gandrud, A. Langford, P. Murphy, R.B. Norton, K.E. Pickering, and B. Ridley. 1987. A ground-based intercomparison of NO, NO_x, NO_y measurement techniques. Journal of Geophysical Research 92:14710-14722.

Ferrell, C.L., and T.G. Jenkins. 1998. Body composition and energy utilization by steers of diverse genotypes fed a high-concentrate diet during the finishing period: 1. Angus, Belgian Blue, Hereford and Piedmontese Sires. Journal of Animal Science 76:637-646.

Findley, R.W. 2000. Juridical action of private parties. Part V of International Encyclopaedia of Laws, 2d ed., M.R. Grossman et al. (eds.). Environment Law—USA. The Hague: Kluwer Law International.

Finlayson-Pitts, B.J., and J.N. Pitts. 2000. Chemistry of the Upper and Lower Atmosphere: Theory, Experiments and Applications. Academic Press.

Fitzpatrick, M. 1987. Emission Control Technologies and Emission Factors for Unpaved Road Fugitive Emissions User's Guide, EPA/625/5-87/022. Washington, D.C.

Flessa, H., and F. Beese. 2000. Laboratory estimates of trace gas emissions following surface application and injection of cattle slurry. Journal of Environmental Quality 29:262-268.

Flessa, H., P. Dorsch, and F. Beese. 1995. Seasonal variation of N_2O and CH_4 fluxes in differently managed arable soils in southern Germany. Journal of Geophysical Research-Atmospheres 100:23115-23124

Flessa, H., P. Dorsch, F. Beese, H. Konig, and A.F. Bouwman. 1996. Influence of cattle wastes on nitrous oxide and methane fluxes in pasture land. Journal of Environmental Quality 25:1366-1370.

Fowler, D. 1999. Experimental designs appropriate for flux determination in terrestrial and aquatic ecosystems. Pp. 100-121 in Approaches to Scaling Trace Gas Fluxes in Ecosystems, A.F. Bouwman (ed.), Elsevier Science.

Freney, J.R. 1967. Sulfur-containing organics. Pp. 229-259 in Soil Biochemistry, A.D. McLaren and G.H. Petersen (eds.). New York: Marcel Dekker.

Fritz, B., C.B. Parnell, Jr., and B.W. Shaw. 2001. Point source particulate dispersion modeling: A summary of findings. Proceedings of the 2001 Beltwide Cotton Production Conferences. National Cotton Council, Memphis, Tenn.

FSA (Farm Service Agency). 2002. Who we are, and what we do. Available on-line at http://www.fsa.usda.gov/pas/about_us/mission.htm [October 2002].

Galloway, J.N., and E.B. Cowling. 2002. Reactive nitrogen and the world: Two hundred years of change. Ambio 31:64-71.

Gao, W. 1995. The vertical change of coefficient *b*, used in the relaxed eddy accumulation method for flux measurement above and within a forest canopy. Atmospheric Environment 29:2339-2347.

Gollehon, N., M. Caswell, M. Ribaudo, R. Kellogg, C. Lander, and D. Letson. 2001. Confined Animal Production and Manure Nutrients. USDA Agriculture Information Bulletin No. 771. Washington, D.C.

Gordy, J. F., 1974. Broilers. In: American Poultry History: 1823-1973, O. A. Hanke, J. L. Skinner and J. H. Florea (eds.). Madison, Wisc.: American Printing and Publishing. Pp. 370-433.

Grelinger, M.A. 1998. Improved emission factors for cattle feedlots. Emission Inventory: Planning for the Future, Proceedings of Air and Waste Management Association, U.S. Environmental Protection Agency Conference 1:515-524 (October 28-30, 1997).

Grelinger, M.A., and A. Page. 1999. Air pollutant emission factors for swine facilities. Pp. 398-408 in Air and Waste Management Conference Proceedings.

Groenestein, C.M., and H.G. VanFaassen. 1996. Volatilization of ammonia, nitrous oxide and nitric oxide in deep-litter systems for fattening pigs. Journal of Agricultural Engineering Research 65:269-274.

Groot Koerkamp, P.W.G., J.H.M. Metz, G.H. Uenk, V.R. Phillips, M.R. Holden, R.W. Sneath, J.L. Short, R.P. White, J. Hartung, and J. Seedorf. 1998. Concentrations and emissions of ammonia in livestock buildings in Northern Europe. Journal of Agricultural Engineering Research 70:79-95.

Grossman, M.R. 1994. Agriculture and the environment in the United States. American Journal of Comparative Law (Suppl.): 291-338.

Grub, W., C.A. Rollo, and J.R. Howes. 1965. Dust Problems in Poultry Environments. Dust and Air Filtration in Animal Shelters (a symposium). American Society of Agricultural Engineers. 338 pp.

Guenther, A.B., and A.J. Hills. 1998. Eddy covariance measurement of isoprene fluxes. Journal of Geophysical Research 103:13145-13152.

Guenther, F.R., W.D Dorko, W.R., Miller, and G.C. Rhoderick. 1996. The NIST traceable reference material program for gas standards. Washington, D.C.: U.S. Department of Commerce, National Institute of Standards and Technology. NIST special publication 260-126.

Hansen, S., J.E. Maehlum, and L.R. Bakken. 1993. N_2O and CH_4 fluxes in soil influenced by fertilization and tractor traffic. Soil Biology Biochemistry 25:621-630.

Harper, L.A., and R.R. Sharpe. 1998. Pp. 1-22 in Ammonia Emissions from Swine Waste Lagoons in the Southeastern U.S. Coastal Plains. Final Report for USDA-ARA Agreement No. 58-6612-7M-022. Division of Air Quality, N.C. Department of Environment and Natural Resources, Raleigh, N.C.

Harper, L.A., R.R. Sharpe, and T.B. Parkin. 2000. Gaseous nitrogen emissions from anaerobic swine lagoons: Ammonia, nitrous oxide and dinitrogen gas. Journal of Environmental Quality 29:1356-1365.

Harrison, R.M., and A.M.N. Kitto. 1990. Field inter-comparison of filter pack and denuder sampling methods for reactive gaseous and particulate pollutants. Atmospheric Environment 24:2633-2640.

Havenstein, G.B., P.R. Ferket, and S.E. Scheideler. 1994. Carcass composition and yield of 1991 vs. 1957 broilers when fed typical 1957 and 1991 broiler diets. Poultry Science 73:1795-1804.

Havenstein, G.B., P.R. Ferket, and M.A. Qureshi. 2002. Growth, feed efficiency and livability of 1957 vs. 2001-type broilers when fed typical 1957 and 2001-type diets. Abstract 369, Poultry Science Association annual meeting, Newark, Del., August 10-14. Poultry Science (Suppl. 1) 81.

Heber, A.J., D.S. Bundy, T.T. Lim, J.Q. Ni, B.L. Haymore, C.A. Diehl, and R.K. Duggirala. 1998. Odor emission rates from swine confinement buildings. Pp. 304-310 in Animal Systems and the Environment. Proceedings, International Conference on Odor, Water Quality, Nutrient Management, and Socieoeconomic Issues, Des Moines, Iowa, July 20-22.

Heber, A.J., T. Lim, J. Ni, D. Kendall, B. Richert, and A.L. Sutton. 2001. Odor, ammonia and hydrogen sulfide emission factors for grow-finish buildings. #99-122. Clive, Iowa: Final Report. National Pork Producers Council.

Henry, R.C., C.H. Spiegelman, and S.L Dattner. 1995. Multivariate receptor modeling of Houston Auto-GC VOC Data. Proceedings APCA Annual Meeting 10:95.

Hinz, T., and S. Linke, 1998. A comprehensive experimental study of aerial pollutants in and emissions from livestock buildings. Part 2: Results. Journal of Agricultural Engineering Research 70:119-129.

Hobbs, P.J., T.H. Misselbrook, M.S. Dhanoa, and K.C. Persaud. 2001. Development of a relationship between olfactory response and major odorants from organic wastes. Journal of the Science of Food and Agriculture 81:188-193.

Hobbs, P.J., T.H. Misselbrook, and T.R. Cumby. 1999. Production and emission of odours and gases from aging pig waste. Journal of Agricultural Engineering Research 72:291-298.

Hobson, P.N. 1990. The treatment of agricultural wastes. Pp. 93-139 in Anaerobic Digestion: A Waste Treatment Technology, A. Wheatley (ed.). Dordrecht: Kluwer Academic Publishers.

Hoeksma, P., N. Verdoes, J. Ooosthoek, and J.A.M. Voermans. 1982. Reduction of ammonia volatilization from pig houses using aerated slurry as recirculation liquid. Livestock Production Science 31:121-132.

Holmen, B.A., T.A. James, L.L. Ashbaugh, and R.G. Flocchini. 2001. Lidar-assisted measurement of PM10 emissions from agricultural tilling in California's San Joaquin Valley. Atmospheric Environment 35:3251-3277.

Howarth, R.W., E.W. Boyer, W.J. Pabich and J.N. Galloway, 2002. Nitrogen use in the United States from 1961-2000 and potential future trends. Ambio 31(2): 88-96.

Huey, N.A., L.A. Broering, G.A. Jutze, and G.W. Guber. 1960. Objective odor pollution control investigations. Journal of Air Pollution Control Association 10:441.

Huff, F.F., and J.R. Angel. 1992. Rainfall Frequency Atlas of the Midwest. Champaign, Ill.: Illinois Water Survey Bulletin 71.

Humenik, F.J. 2001. History of Lagoons and Animal Waste Management in North Carolina. In Proceedings (CD) of the International Symposium Addressing Animal Production and Environmental Issues. College of Agriculture and Life Sciences, North Carolina State University, Raleigh.

Hurrell, R.F., R. Deutsch, and P.A. Finot. 1980. Effect of ultra-high temperature steam injection on sulfur-containing amino acids of skim milk. Journal of Dairy Science 2:298-300.

Hutcheson, J.P., D.E. Johnson, C.L. Gerken, J.B. Morgan, and J.D. Tatum. 1997. Anabolic implant effects on visceral organ mass, chemical body composition, and estimated energetic efficiency in cloned (genetically identical) beef steers. Journal of Animal Science 75:2620-2626.

Hutchinson , G.L., A.R. Mosier, and C.E. Andre. 1982. Ammonia and amine emissions from a large cattle feedlot. Journal of Environmental Quality 11:288-293.

IPCC (Intergovernmental Panel on Climate Change). 2001. Climate Change 2001: The Scientific Basis. Contributions of Working Group I to the Third Assessment Report of the Intergovernmental Panel on Climate Change, J.T. Houghton, D.J. Griggs, M. Noguer, P.J. van der Linden, X. Dai, K. Maskell, and C.A. Johnson (eds.) Cambridge, U.K.: Cambridge University Press. 881 pp.

IPCC (Intergovernmental Panel on Climate Change). 2000. Good Practice Guidance and Uncertainty Management in National Greenhouse Gas Inventories.

Jacobson, L. 1999. Odor and gas emissions from animal manure storage units and buildings. Paper presented at ASAE Annual Meeting, Toronto, Canada.

Jarvis, S.C. 1997. Emission processes and their interactions in grassland soils. Pp. 1-17 in Gaseous Nitrogen Emissions from Grasslands, S.C. Jarvis and B.F. Pain (eds.). New York: CAB International.

Jarvis S.C., and B.F. Pain. 1994. Greenhouse-gas emissions from intensive livestock systems: their estimation and technologies for reduction. Climatic Change 27:27-38.

Jiang, J.K., and J.R. Sands. 1998. Report on odour emissions from poultry farms in western Australia. Principal Technical Report, Odour Research Laboratory, Centre for Water and Waste Technology, School of Civil and Environmental Engineering, University of New South Wales, Sydney, Australia.

Johnson, D.E., and G.M. Ward. 1996. Estimates of animal methane emissions. Environmental Monitoring and Assessment 42:133-141.

Johnson, D.E., K.A. Johnson, G.M. Ward and M.E. Branine. 2000. Ruminants and Other animals. Pp. 112-133 in Atmospheric Methane: Its Role in the Global Environment, 2nd ed., M.A.K. Khalil (ed). Berlin: Springer-Verlag.

Johnson, K.A., and D.E. Johnson. 1995. Methane emissions from cattle. Journal of Animal Science 73:2483-2492.

Jonker, J.S., R.A. Kohn, and R.A. Erdman. 1999. Milk urea nitrogen target concentrations for lactating dairy cattle fed according to National Research Council recommendations. Journal of Dairy Science 82:1261-1273.

Jonker, J.S., R.A. Kohn, and J. High. 2002. Dairy herd management practices that impact nitrogen utilization efficiency. Journal of Dairy Science 85:1218-1226.

Jungbluth, T., E. Hartung, and G. Brose. 2001. Greenhouse gas emissions from animal houses and manure stores. Nutrient Cycling in Agroecosystems 60:122-145.

Katanbaf, M.N., E.A. Dunnington, and P.B. Siegel. 1989. Restricted feeding in early and late-feathering chickens. 3. Organ size and carcass composition. Poultry Science 68:359-368.

Kellogg, R.L. 2002. Profile of farms with livestock in the United States: A statistical summary. Natural Resources Conservation Service, U.S. Department of Agriculture. Washington, D.C.

King, J.J. 1995. The Environmental Dictionary, 3rd ed. New York: John Wiley & Sons.

Kline, R.D., V.W. Hays, and G.L. Cromwell. 1971. Effects of copper, molybdenum and sulfate on performance, hematology and copper stores of pigs and lambs. Journal of Animal Science 33:771-770.

Klopfenstein, T., R. Angel, G.L. Cromwell, G.E. Erickson, D.G. Fox, C. Parsons, L.D. Satter, and A.L. Sutton. 2002. Animal diet modification to decrease the potential for nitrogen and phosphorus pollution. Issue Paper 21. Council for Agriculture, Science, and Technology. Ames, Iowa.

Kohn, R.A., Z. Dou, J.D. Ferguson, and R.C. Boston. 1997. A sensitivity analysis of nitrogen losses from dairy farms. Journal of Environmental Management 50:417-428.

Kornegay, E.T. (ed.). 1996. Nutrient Management of Food Animals to Enhance and Protect the Environment. Boca Raton, Fla.: CRC Press.

Kroodsma, W., R. Scholtens, and J. Huis in't Veld. 1988. Ammonia emissions from poultry housing systems volatile emissions from livestock farming and sewage operations. Proceedings of CIGR Seminar Storing, Handling and Spreading of Manure and Municipal Waste 2:7.1-7.13.

Lacey, R.E. 1998. Characterization of Swine Production Odors by an Electronic Nose. Texas Agricultural Experiment Station, College Station, Tex.

Lacy, M.P., and M. Czarick. 1992. Tunnel-ventilated broiler houses: Broiler performance and operating costs. Journal of Applied Poultry Research 1:104-109.

Lamb, B., H. Westberg, G. Allwine, and T. Quarles. 1985. Biogenic hydrocarbon emissions from deciduous and coniferous trees in the United States. Journal of Geophysical Research 90:2380-2390.

Lapitan, R.L., R. Wanninkhof, and A.R. Mosier. 1999. Methods for stable gas flux determination in aquatic and terrestrial systems. Pp. 29-66 in Approaches to Scaling Trace Gas Fluxes in Ecosystems, A.F. Bouwman (ed.). Elsevier Science.

Leng, R.A. 1993. Quantitative ruminant nutrition—A green science. Australian Journal of Agricultural Research 44:363-380.

Lessard, R., P. Rochette, E.G. Gregorich, E. Pattey, and R.L. Desjardins. 1996. Nitrous oxide fluxes from manure-amended soil under maize. Journal of Environmental Quality 25:1371-1377.

Leuning, R., S.K. Baker, I.M. Jamie, C.H. Hsu, L. Klein, O.T. Denmead, and D.W.T. Griffith. 1999. Methane emission from free-ranging sheep: A comparison of two measurement methods. Atmospheric Environment 33:1357-1365.

Li, C.S., V. Narayanan, and R.C. Harriss. 1996. Model estimates of nitrous oxide emissions from agricultural lands in the United States. Global Biogeochemical Cycles 10:297-306.

Liang, Z.S., P.W. Westerman, and J. Arogo. 2002. Modeling ammonia emission from swine anaerobic lagoons. Transactions of the American Society of Agricultural Engineers 45:787-798.

Lim, T.T., A.J. Heber, J.Q. Ni, A.L. Sutton, and D.T. Kelly. 2001. Characteristics and emission rates of odor from commercial swine nurseries. Transactions of the ASAE 44:1275–1282.

Lorimor, J., K. Kohl, R. Killorn, B. Lotz, and P. Miller. 1999. Dry Manure Applicator Certification Study Guide. Iowa State University, University Extension.

Lorimor, J., C. Fulhage, R. Zhang, T. Funk, R. Sheffield, D. C. Sheppard, and G.L. Newton. 2001. Chapter 8: Manure management strategies/technologies in [CD-ROM] National Center for Manure and Animal Waste Management White Papers. Ames, Iowa: Midwest Planning Service.

MacIntyre, S., R. Wanninkhof, and J.P. Chanton. 1995. Trace gas exchange across the air-water interface in freshwater and coastal marine environments. Pp. 52-97 in Biogenic Trace Gases: Measuring Emissions from Soil and Water, P.A. Matson and R.C. Harriss (eds.). Methods in Ecology. Cambridge Mass: Blackwell Science, Ltd.

Mackay-Sim, A. (ed.). 1992. Electronic odour detection—problems and possibilities. Pp. 57-61 in Odour Update '92: Proceedings Workshop on Agricultural Odours. MRC Report No. DAQ 64/24 ed. Toowoomba, Queensland: Department of Primary Industries.

Mahan, D.C., and R.G. Shields, Jr. 1998. Macro and micro mineral composition of pigs from birth to 145 kilograms of body weight. Journal of Animal Science 766:506-512.

Malone, Linda A. 2001. Environmental Regulation of Land Use. St Paul, Minn.: West Group.

Manchester, A., and D. Blayney. 1997. Structure of Dairy Markets: Past, Present, and Future. ERS AER 757, Economic Research Service, U.S. Department of Agriculture. Washington, D.C.

Martinez, S. 2002. Vertical Coordination of Marketing Systems: Lessons from the Poultry Egg and Pork Industries. AER No. 807, Economic Research Service, U.S. Department of Agriculture. Washington, D.C.

Masarie, K.A., R.L. Langenfelds, C.E. Allison, T.J. Conway, E.J. Dlugokencky, R.J. Francey, P.C. Novelli, L.P. Steele, P.P. Tans, B. Vaughn, and J.W.C. White. 2001. The NOAA/CSIRO flask-air intercomparison program: A strategy for directly assessing consistency among atmospheric measurements derived from independent laboratories. Journal of Geophysical Research 106:20445-20464.

McCaughey, W.P., K. Wittenberg, and D, Corrigan. 1999. Impact of pasture type on methane production by lactating beef cows. Canadian Journal of Animal Science 79:221-226.

Meister, M., B. W. Shaw, B. Fritz, and C. B. Parnell, Jr. 2001. Dispersion modeling of ground level area sources. Proceedings of the 2000 Beltwide Cotton Production Conferences. National Cotton Council. Memphis, Tenn.

Miner, J.R. 1975. Management of odors associated with livestock production. In Managing Livestock Wastes. Proceedings of 3rd International Symposium on Livestock Wastes. St. Joseph, Mich: American Society of Agricultural Engineers.

Miner, J.R., and L.A. Licht. 1981. Fabric swatches as an aid in livestock manure odor evaluations. Pp. 297-301 in Livestock Wastes: A Renewable Resource. Proceedings of the 4th International Symposium on Livestock Wastes—1980. American Society of Agricultural Engineers, St. Joseph, Mich.

Miner, J.R., and R.C. Stroh. 1976. Controlling feedlot surface odor emission rates by application of commercial products. Transactions of the American Society of Agricultural Engineers 19(3):533-538.

Miner, J.R., F.J. Humenik, and M.R. Overcash. 2000. Managing Livestock Wastes to Preserve Environmental Quality. Ames: Iowa State University Press.

Moe, P.W., and H. F. Tyrrell. 1979. Methane production in dairy cows. Journal of Dairy Science 62:1583-1586.

Monteny G.J., and J.W. Erisman. 1998. Ammonia emission from dairy cow buildings: A review of measurement techniques, influencing factors and possibilities for reduction. Netherlands Journal of Agricultural Science 46:225-247.

Moore, P.A., Jr., B.C. Joern, D.R. Edwards, C.W. Wood, and T.C. Daniel. 2001. Chapter 14: Effects of manure amendments on environmental and production problems. In [CD-ROM] National Center for Manure and Animal Waste Management White Papers. Ames, Iowa : Midwest Planning Service.

Mosier, A., C. Kroeze, C. Nevison, O. Oenema, S. Seitzinger, and O. van Cleemput. 1998. Closing the global N_2O budget: Nitrous oxide emissions through the agricultural nitrogen cycle—OECD/IPCC/IEA phase II development of IPCC guidelines for national greenhouse gas inventory methodology. Nutrient Cycling in Agroecosystems 52:225-248.

Mosier, A.R., J.M. Duxbury, and J.R. Freney. 1996. Nitrous oxide emissions from agricultural fields: assessment, measurement and mitigation. Plant Soil 181:95-108.

Mount, G. H., B. Rumburg, J. Havig, B. Lamb, H. Westberg, D. Yonge, K. Johnson, and R. Kincaid. 2002. Measurement of atmospheric ammonia at a dairy using differential optical absorption spectroscopy in the mid-ultraviolet. Atmospheric Environment 36:1799-1810.

Muck, R.E., and T.S. Steenhuis. 1982. Nitrogen losses from manure storages. Agricultural Wastes 4:41-48.

Müller, C., R.R. Sherlock, K.C. Cameron, and J.R.F. Barringer. 1997. Application of a mechanistic model to calculate nitrous oxide emissions at a national scale. Pp. 339-349 in Gaseous Nitrogen Emissions from Grasslands, S.C. Jarvis and B.F. Pain (eds.). New York: CAB International.

MWPS (Midwest Plan Service). 1985. Livestock Waste Facilities Handbook (MWPS-18). Ames: Iowa State University

MWPS (Midwest Plan Service). 1989. Natural Ventilating Systems for Livestock Housing (MWPS-33). Ames: Iowa State University

MWPS (Midwest Plan Service). 1990. Mechanical Ventilating Systems for Livestock Housing (MWPS-32). Ames: Iowa State University

MWPS (Midwest Plan Service). 1992. Livestock Waste Facilities Handbook (MWPS-18). Ames: Iowa State University.

NAPAP (National Acid Precipitation Assessment Program). 1990. Acidic Deposition: State-of-Science and State-of-Technology, Vols. I-IV, P.M. Irving (ed.). Washington, D.C.: National Acid Precipitation Assessment Program.

NBS (National Bureau of Standards). 1975. Catalog of NBS Standard Reference Materials, 1975-76 ed., NBS special publication no. 260. Washington, D.C.: U.S. Department of Commerce, National Bureau of Standards.

Ni, J.-Q., A.J. Heber, T.T. Lim, and C.A. Diehl. 2002a. Continuous measurement of hydrogen sulfide emission from two large swine finishing buildings. Journal of Agricultural Science 138:227-236.

Ni, J.-Q., A.J. Heber, C.A. Diehl, T.T. Lim, R.K. Duggirala, and B.L. Haymore. 2002b. Characteristics of hydrogen sulfide concentrations in two mechanically ventilated swine buildings. Canadian Biosystems Engineering 44:11-19.

Ni, J.-Q., A.J. Heber, C.A. Diehl, T.T. Lim, R.K. Duggirala, and B.L. Haymore. 2002c. Summertime concentrations and emissions of hydrogen sulfide at a mechanically-ventilated swine finishing building. Transactions of ASAE 45:193-199.

Ni, J.Q., A.J. Heber, T.T. Lim, C.A. Diehl, R.K. Duggirala, and B.L. Haymore. 2002d. Hydrogen sulfide emission from two large pig-finishing buildings with long-term high-frequency measurements. Journal of Agricultural Science 138:227-236.

Ni, J.Q., A.J. Heber, C.A. Diehl, T.T. Lim, R.K. Duggirala, and B.L. Haymore. 2000. Burst releases of hydrogen sulfide in mechanically ventilated swine buildings. Odors/VOC Emissions Conference, Cincinnati, Oh., April 17-19.

North Carolina Department of Environmental and Natural Resources—Division of Air Quality. 1999. Status Report on Emissions and Deposition of Atmospheric Nitrogen Compounds from Animal Production in North Carolina.

NRC (National Research Council). 1992. Policy Implications of Greenhouse Warming: Mitigation, Adaptation, and Science Base. Washington, D.C.: National Academy Press.

NRC (National Research Council). 1994. Nutrient Requirements of Poultry, 9th revised ed. Washington, D.C.: National Academy Press.

NRC (National Research Council). 1997. Biodiversity II: Understanding and Protecting Our Biological Resources. Washington, D.C.: National Academy Press.

NRC (National Research Council). 1998a. Nutrient Requirements of Swine, 10th revised ed. Washington, D.C.: National Academy Press.

NRC (National Research Council). 1998b. Research Priorities for Airborne Particulate Matter: I. Immediate Priorities and a Long-Range Research Portfolio. Washington, D.C.: National Academy Press.

NRC (National Research Council). 2000. Nutrient Requirements of Beef Cattle, 7th revised ed. Washington, D.C.: National Academy Press.

NRC (National Research Council). 2001a. Nutrient Requirements of Dairy Cattle, 7th revised ed. Washington, D.C.: National Academy Press.

NRC (National Research Council). 2001b. Climate Change Science: An Analysis of Some Key Questions. Washington, D.C.: National Academy Press.

NRC (National Research Council). 2002a. The Scientific Basis for Estimating Air Emissions from Animal Feeding Operations. Washington, D.C.: National Academy Press.

NRC (National Research Council). 2002b. The Airliner Cabin Environment and Health of Passengers and Crew. Washington, D.C.: National Academy Press.

NRC (National Research Council). 2002c. Biosolids Applied to Land: Advancing Standards and Practices. Washington, D.C.: National Academy Press.

NRC (National Research Council). 2002d. The Ongoing Challenges of Managing Carbon Monoxide Pollution in Fairbanks, Alaska. Washington, D.C.: National Academy Press.

NRCS (Natural Resources Conservation Service). 2001. Accomplishments Report FY 2001. Available on-line at http://www.nrcs.usda.gov/about/spa/accomplishments.pdf [October 2002].

NRCS (Natural Resources Conservation Service). 2002a. Animal Feeding Operations. Available on-line at http://www.nrcs.usda.gov/programs/afo/ [October 2002].

NRCS. (Natural Resources Conservation Service). 2002b. Environmental Quality Incentives Program. Fact Sheet.

NRCS (Natural Resources Conservation Service). 2002c. Initial performance plan for FY2003, and revised plan for RY2002. Available on-line at http://www.nrcs.usda.gov/about/spa/perfplans.pdf.

NRCS (Natural Resources Conservation Service). 2002d. National Planning Procedures Handbook. Draft Comprehensive Nutrient Management Planning Technical Guidance. Available on-line at http://www.nrcs.usda.gov/programs/afo/cnmp_guide_index.html.

Oenema, O., A. Bannink, S.G. Sommer, and L. Velthof. 2001. Gaseous nitrogen emissions from livestock farming systems. Pp. 255-289 in Nitrogen in the Environment: Sources, Problems, and Management, R.F. Follett, and J.L. Hatfield (eds.). Elsevier. 520 pp.

Ogink, N.W.M., C. van ter Beek, and J.V. Karenbeek. 1997. Odor Emission from Traditional and Low-Emitting Swine Housing Systems: Emission Levels and Their Accuracy. ASAE Paper No. 97-4036. St. Joseph, Mich.: American Society of Agricultural Engineers.

Olivier, J.G.J., A.F. Bouwman, K.W. van der Hoek, and J.J.M. Berdowski. 1998. Global Air Emission Inventories for Anthropogenic Sources of NO_x, NH_3, and N_2O in 1990. Environmental Pollution 102:135-148.

Olivier J.G.J., L.J. Brandes, J.A.H.W. Peters, and P.W.H.G. Coenen. 2002. Greenhouse Gas Emissions in the Netherlands 1990-2000. National Inventory Report 2002. RIVM Rapport 773201006. Rijksinstituut voor Volkagezondheld en Milieu. 150 pp.

Ollinger, S.V., J.D. Aber, P.B. Reich, and R.J. Freuder. 2002. Interactive effects of nitrogen deposition, tropospheric ozone, elevated CO_2, and land use history on the carbon dynamics of northern hardwood forests. Global Change Biology 8:1-18.

Owens, M.A., C.C. Davis, and R.R. Dickerson. 1999. A new photothermal interferometer for ammonia detection. Analytical Chemistry 71:1391-1399.

Pace, T.G. 1985. Receptor methods for source apportionment: Real world issues and applications. Journal of the Air Pollution Control Association 1149-1153.

Pain, B.F., T.H. Misselbrook, C.R. Clarkson, and Y.J. Rees. 1990. Odor and ammonia emissions following the spreading of anaerobically-digested pig slurry on grassland. Biological Wastes 34:259-267.

Parnell, S.E., B. Lesikar, J.L. Sweeten, and R.E. Lacey. 1994. Determination of an emission factor for cattle feedyards by applying dispersion modeling. Paper 94-4042/94-4082 (Summer). American Society of Agricultural Engineers Kansas City, Missouri. 15 pp.

Paul, J.W., and E.G. Beauchamp. 1993. Nitrogen availability for corn in soils amended with urea, cattle slurry, and solid and composted manures. Canadian Journal of Soil Science. 73:253-266.

Penner, J.E., M. Andreae, A. Annegarn, L. Barrie, J. Feichter, D. Hegg, A. Jayaraman, R. Leaitch, D. Murphy, J. Nganga, and G. Pitari. 2001. Aerosols, their direct and indirect effects. Pp. 289-348 in Climate Change 2001: The Scientific Basis. Contributions of Working Group I to the Third Assessment Report of the Intergovernmental Panel on Climate Change, J.T. Houghton, D.J. Griggs, M. Noguer, P.J. van der Linden, X. Dai, K. Maskell, C.A. Johnson (eds.). Cambridge, U.K.: Cambridge University Press. 881 pp.

Peters, J.A., and T.R. Blackwood. 1977. Source Assessment: Beef Cattle Feedlots. EPA-600/ 2-77-107. Research Triangle Park, N.C. U.S. Environmental Protection Agency. 101 pp.

Petersen, S.O. 1999. Nitrous oxide emissions from manure and inorganic fertilizers applied to spring barley. Journal of Environmental Quality 28:1610-1618.

Phillips, V.R., D.S. Lee, R. Scholtens, J.A. Garland, and R.W. Sneath. 2001. A review of methods for measuring emission rates of ammonia from livestock buildings and slurry or manure stores. Part 2: Monitoring flux rates, concentrations and airflow rates. Journal of Agricultural Engineering Research 78(1):1-14.

Powers, W., and R. Koelsch. 2002. National standards for estimating manure nutrient excretion based on animal feed program. Journal of Animal Science (Suppl. 1) 80:137 (abstr).

Powers, W.J., H.H. Van Horn, A.C. Wilkie, C.J. Wilcox, and R.A. Nordstedt. 1999. Effects of anaerobic digestion and additives to effluent or cattle feed on odor and odorant concentration. Journal of Animal Science 77:1412-1421.

Powers, W.J., T. van Kempen, D.S. Bundy, A. Sutton, and S.J. Hoff. 2000. Objective measurement of odors using gas chromatography/mass spectrometry and instrumental technologies. Pp. 163-169 in Proceedings of 2nd International Conference on Air Pollution from Agricultural Operations. St. Joseph, Mich.: American Society of Agricultural Engineers.

Prather, M., D. Ehalt, F. Dentener, R. Derwent, Dlugokencky, E. Holland, I. Isaksen, J. Katima, Kirchhoff, P. Matson, P. Midgley, and M. Wang. 2001. Atmospheric chemistry and greenhouse gases. In Climate Change 2001: The Scientific Basis. Contributions of Working Group I to the Third Assessment Report of the Intergovernmental Panel on Climate, J.T. Houghton, D.J. Griggs, M. Noguer, P.J. van der Linden, X. Dai, K. Maskell, and C.A. Johnson (eds.). Cambridge, U.K. and New York: Cambridge University Press. 881 pp.

Ramadan, Z., X.-H. Song, P.K. Hopke. 2000. Identification of sources of Phoenix aerosol by positive matrix factorization. Journal of Air Waste Management Association 50:1308-1320.

Redwine, J.S., and R.E. Lacey. 2000. A summary of state-by-state regulation of livestock odor. Second International Conference on Air Pollution from Agricultural Operations, Des Moines, Iowa: American Society of Agricultural Engineers.

Rinne, H., A. Guenther, C. Warneke, J. deGouw, and S. Luxembourg. 2001. Disjunct eddy covariance technique for trace gas flux measurements. Geophysical Research Letters 28:3139-3142.

Riviere, J., J.C. Subtil, and G. Catroux. 1974. Etude de l'evolution physico-chiminique et microbiologique du lisier de porcs pendant le stockage anaerobie. Annales Agronomiques 15:383-401.

Robertson, G.P., E.A. Eldor, R.R. Harwood. 2000. Greenhouse gases in intensive agriculture: Contributions of individual gases to the radiative forcing of the atmosphere. Science 289:1922-1925.

Rochette, P., E. van Bochove, D. Prevost, D.A. Angers, D. Cote, and N. Bertrand. 2000. Soil carbon and nitrogen dynamics following application of pig slurry for the 19th consecutive year: II. Nitrous oxide fluxes and mineral nitrogen. Soil Science Society of America Journal 64:1396-1403.

Roka, F. 1993. An economic analysis of joint production relationships between pork, and swine manure. Unpublished Ph.D. dissertation, Department of Agricultural, and Resource Economics, Raleigh: North Carolina State University.

Royal Society of Chemistry. 1991. The Composition of Foods, 5th ed. London.

Ruhl, J.B. 2000. Farms, their environmental harms, and environmental law. Ecology Law Quarterly 27(2): 263-350.

Safley, L.M, Jr. 1992. Low temperature digestion of dairy, and swine manure. ASAE Paper No. 92-6612 presented at the International Winter Meeting of the American Society of Agricultural Engineers, Nashville, Tenn., December 15-18.

Safley, L.M., Jr., and P.W. Westerman. 1988. Biogas production from anaerobic lagoons. Biological Wastes 23:181-193.

Safley, L.M., Jr., and P.W. Westerman. 1990. Psychrophilic anaerobic digestion of animal manure: Proposed design methodology. Biological Wastes 34:133-148.

Santoso, U., K. Tanaka, and S. Ohtani. 1995. Early skip-a-day feeding of female broiler chicks fed high-protein realimentation diets. Performance, and body composition. Poultry Science 74:494-501.

Sawyer, G., 1971. The Agribusiness Poultry Industry: A History of Its Development. New York: Exposition Press.

Schaefer, J. 1977. Sampling, characterization, and analysis of malodor. Agricultural Environment 3:121-128.

Schiffman, S.S., and C.M. Williams. 1999. Evaluation of swine odor control products using human odor panels. Pp. 110-118 in Animal Waste Management Symposium. Raleigh: North Carolina State University.

Schiffman, S.S., J.L. Bennett, and J.H. Raymer. 2001. Quantification of odors, and odorants from swine operations in North Carolina. Agricultural Forest Meteorology 108:213-240.

Schiffman, S.S., E.A. Sattely Miller, M.S. Suggs, and B.G. Graham. 1995. The effect of environmental odors emanating from commercial swine operations on the mood of nearby residents. Brain Research Bulletin 37:369-375.

Schmidt, D. 2000. Odor, hydrogen sulfide, and ammonia emissions from composting of caged layer manure. Final report to Broiler and Egg Association of Minnesota.

Schnoor, J.L., P.S. Thorne, and W. Powers. 2002. Fate and transport of air pollutants from CAFOs. Pp. 86-100 in Iowa Concentrated Animal Feeding Operation Air Quality Study. Environmental Health Sciences Research Center, University of Iowa. Available on-line at http://www.public-health.uiowa.edu/ehsrc/CAFOstudy.htm.

Schwartz, J. 1994. Air pollution and daily mortality: A review and meta analysis. Environmental Research 64:36-52.

Seinfeld, J.H., and S.N. Pandis. 1998. Atmospheric Chemistry and Physics in Air Pollution and Climate Change. New York: John Wiley & Sons.

Shaw, W.J., C.W. Spicer, and D.V. Kenny. 1998. Eddy correlation fluxes of trace gases using a tandem mass spectrometer. Atmospheric Environment 32:2887-2898.

Siopes, T.D. 1999. Intermittent lighting increases egg weight and facilitates early photostimulation of turkey breeder hens. Poultry Science 78:1040-1043.

Skiba, U., D. Fowler, and K.A. Smith. 1997. Nitric oxide emissions from agricultural soils in temperate and tropical climates: Sources, controls and mitigation options. Nutrient Cycling in Agroecosystems 48:139-153.

Slemr, F., and W. Seiler. 1984. Field emissions of NO and NO_2 from fertilized and unfertilized soils. Journal of Atmospheric Chemistry 2:1-24.

Smil, V. 1999. Nitrogen in crop production: An account of global flows. Global Biogeochemical Cycles 13:647-662.

Smil, V. 2001. Enriching the Earth. Cambridge, Mass.: MIT Press

Smith, K.A., I.P. McTaggart, and H. Tsuruta. 1997. Emissions of N_2O and NO associated with nitrogen fertilization in intensive agriculture, and the potential for mitigation. Soil Use and Management 13:296-304.

Speece, R.E. 1996. Anaerobic Biotechnology for Industrial Wastewater. Nashville, Tenn.: Archae Press.

Spoelstra, S.F. 1980. Origin of objectionable odorous components in piggery wastes and the possibility of applying indicator components for studying odor development. Agricultural Environment 5:241-260.

Stedman, D.H., and R.E. Schetter. 1983. The global budget of atmospheric nitrogen species. Pp. 411-454 in Trace Atmospheric Constituents, S.E. Schwartz (ed.). New York: John Wiley & Sons.

St-Pierre, N.R., and C.S. Thraen. 1999. Animal grouping strategies, sources of variation, and economic factors affecting nutrient balance on dairy farms. Journal of Animal Science 77 (Suppl. 2):73-83.

Sutton, A.L., K.B. Kephart, J.A. Patterson, R. Mumma, D.T. Kelly, E. Bogus, D.D. Jones, and A.J. Heber. 1996. Manipulating swine diets to reduce ammonia and odor emissions. Pp. 445-452 in International Conference on Air Pollution from Agricultural Operations, Kansas City, Mo., February 7-9.

Sweeten, J.M., and J.R. Miner. 1993. Odor intensities at cattle feedlots in nuisance litigation. Bioresource Technology 45:177-188.

Sweeten, J.M., D.L. Reddell, L. Schake, and B. Garner. 1977. Odor intensities at cattle feedlots. Transactions of the American Society of Agricultural Engineers 20:502-508.

Sweeten, J.M., D.L. Reddell, A.R. McFarland, R.O. Gauntt, and J.E. Sorel. 1983. Field measurement of ambient odors with a butanol olfactometer. Transactions of the American Society of Agricultural Engineers 26:1206-1216.

Sweeten, J.M., R.C. Childers, J.S. Cochran, and R. Bowler. 1991. Odor control from poultry manure composting plant using a soil filter. Applied Engineering in Agriculture 7:439-449.

Sweeten, J.M., C.B. Parnell, B.W. Shaw, and B.W. Auvermann. 1998. Particle size distributions of cattle feedlot dust emissions. Transactions of American Society of Agricultural Engineering 41:1477-1478.

Sweeten, J. M., C. B. Parnell, B. W. Auvermann, B. W. Shaw, and R. E. Lacey. 2000. Livestock feedlots. Pp. 488-496 in Air and Waste Management Association's Air Pollution Engineering Manual, W. T. Davis (ed.). New York: John Wiley & Sons.

Sweeten, J.M., L.D. Jocobson, A.J. Heber, D.R. Schmidt, J.C. Lorimor, P.W. Westerman, J.R. Miner, R. Zhang, C.M. Williams, and B.W. Auverman. 2001. Chapter 1 in Odor Mitigation for Concentrated Animal Feeding Operations [CD ROM]. National Center for Manure and Animal Waste Management (USDA), Ames, Iowa.

Takai, H., S. Pedersen, J.O. Johnsen, J.H.M. Metz, P.W.G. Groot Koerkamp, G.H. Uenk, V.R. Phillips, M.R. Holden, R.W. Sneath, J.L. Short, R.P. White, J. Hurtung, J. Seedorf, M. Schroder, K.H. Linkert, and C.M. Wathes. 1998. Concentrations and emissions of airborne dust in livestock buildings in Northern Europe. Journal of Agricultural Engineering Research 70:59-70.

Tate, R.L. 1995. Soil Microbiology. New York: John Wiley & Sons.

Thompson, R.B., J.C. Ryden, and D.R. Lockyer. 1987. Fate of nitrogen in cattle slurry following surface application or injection to grassland. Journal of Soil Science 38:689-700.

Thu, K., K. Donham, R. Ziegenhorn, S.J. Reynolds, P.S. Thorne, P. Subramanian, W. Whitten, and J. Stookesberry. 1997. A control study of health and quality of life of residents living in the vicinity of large scale swine production facilities. Journal of Agricultural Health Safety 3:13-26.

Todd, L.A., M. Ramanathan, K. Mottus, R. Katz, A. Dodson, and G. Mihan. 2001. Measuring chemical emissions using open-path Fourier transform infrared (OP-FTIR) spectroscopy and computer-assisted tomography. Atmospheric Environment 35:1937-1947.

USDA (U.S. Department of Agriculture). 1941. Factors for converting percentages of nitrogen in foods and feeds into percentages of proteins. USDA Circular No. 183. Washington, D.C.

USDA (U.S. Department of Agriculture). 1996a. Part III: Reference of 1996 Dairy Health and Health Management. Fort Collins, Colo.

USDA (U.S. Department of Agriculture). 1996b. World Agricultural Supply and Demand Estimates. Office of the Chief Economist: 321. Available on-line at http://usda.mannlib.cornell.edu/usda/reports/waobr/wasde-bb/.

USDA (U.S. Department of Agriculture). 1997a. Agricultural Atlas of the United States. USDA National Agricultural Statistics Service, Washington, D.C. Available on-line at www.nass.usda.gov/census/census97/atlas97/index.htm.

USDA (U.S. Department of Agriculture). 1997b. World Agricultural Supply and Demand Estimates. Office of the Chief Economist: 333. Available on-line at http://usda.mannlib.cornell.edu/usda/reports/waobr/wasde-bb/.

USDA (U.S. Department of Agriculture). 1998. World Agricultural Supply and Demand Estimates. Office of the Chief Economist: 345. Available on-line at http://usda.mannlib.cornell.edu/usda/reports/waobr/wasde-bb/.

USDA (U.S. Department of Agriculture). 1999a. World Agricultural Supply and Demand Estimates. Office of the Chief Economist: 357. Available on-line at http://usda.mannlib.cornell.edu/usda/reports/waobr/wasde-bb/.

USDA (U.S. Department of Agriculture). 1999b. Dairy Yearbook. Stock #89032. Economic Research Service, Washington D.C. Available on-line at http://jan.mannlib.cornell.edu/data-sets/livestock/89032/.

USDA (U.S. Department of Agriculture). 1999c. US Census of Agriculture, U.S. Summary and State Data, Vol.1 Geographic Area Series, Part 51. AC97-A-51, USDA National Agricultural Statistics Service, Washington, D.C. Available on-line at http://www.nass.usda.gov/census/census97/volume1/us-51/us1_21.pdf.

USDA (U.S. Department of Agriculture). 2000a. Feed Situation and Outlook Yearbook. Economics Research Service, FDS-2000. Available on-line at http://usda.mannlib.cornell.edu/reports/erssor/field/fds-bby/.

USDA (U.S. Department of Agriculture). 2000b. Confined Livestock Air Quality Subcommittee, J.M. Sweeten, Chair. Air Quality Research and Technology Transfer Programs for Concentrated Animal Feeding Operations. USDA Agricultural Air Quality Task Force (AAQTF) Meeting, Washington, D.C.

USDA (U.S. Department of Agriculture). 2001. Hogs and Pigs Quarterly. National Agricultural Statistics Service, Washington, D.C. Available on-line at http://usda.mannlib.cornell.edu/reports/nassr/livestock/php-bb/2001/hgpg1201.pdf.

USDA (U.S. Department of Agriculture). 2002a. National Daily Direct Prior Day Hog Report- Slaughtered Swine (LM_HG201, October 22). Des Moines, Iowa: Agricultural Marketing Service.

USDA (U.S. Department of Agriculture). 2002b. Food and Nutrition Information Center, National Agricultural Library. Available on-line at http://www.nal.usda.gov/fnic.

USDA. (U.S. Department of Agriculture). 2002c. Livestock, Dairy and Poultry Situation and Outlook LDP-M-100. Economic Research Service, Washington D.C. Available on-line at http://jan.mannlib.cornell.edu/reports/erssor/livestock/ldp-mbb/2002/ldp-m100f.pdf.

USDA (U.S. Department of Agriculture). 2002d. NASS Trends, USDA National Agricultural Statistics Service, Washington D.C. Available on-line at http://www.usda.gov/nass/pubs/trends/livestockdairy.htm.

USDA (U.S. Department of Agriculture). 2002e. NASS Cattle. Washington, D.C. Available on-line at http://usda.mannlib.cornell.edu/reports/nassr/livestock/pct-bb/catl0202.pdf.

USDA (U.S. Department of Agriculture) and EPA (U.S. Environmental Protection Agency). 1999. Unified National Strategy for Animal Feeding Operations. Available on-line at http://cfpub.epa.gov/npdes/afo/ustrategy.cfm?program_id=7.

U.S. Department of Commerce, Bureau of the Census. 2000. Section 23: Agriculture. Pp. 663-684 in Statistical Abstract of the United States. Available on-line at http://www.census.gov/prod/2001pubs/statab/sec23.pdf.

van Aardenne, J.A., F.J. Dentener, C.G.M. Klijn Goldewijk, J. Lelieveld, and J.G.J. Olivier. 2001. A 1°-1° resolution dataset of historical anthropogenic trace gas emissions for the period 1890-1990. Global Biogeochemical Cycles 15:909.

Van der Hel, W., M.W.A. Verstegen, L. Pijls, and M. Van Kampen. 1992. Effect of two-day temperature exposure of neonatal broiler chicks on growth performance and body composition during two weeks at normal conditions. Poultry Science 71:2014-2021.

Verdoes, N. and N.W.M. Ogink. 1997. Odour emission from pig houses with low ammonia emission. Pp. 317-325 in Proceedings of the International symposium on Ammonia and Odour Control from Animal Production Facilities, J.A.M. Voermans, and G. Moteny (eds.). Vinkeloord, The Netherlands.

Wagner-Riddle, C., G.W. Thurtell, G.K. Kidd, E.G. Beauchamp, and R. Sweetman. 1997. Estimates of nitrous oxide emissions from agricultural fields over 28 months. Canadian Journal of Soil Science 77:135-144.

Wathes, C.M., V.R. Phillips, M.R. Holden, R.W. Sneath, J.L. Short, R.P. White, J. Hartung, J. Seedorf, M. Schroder, K.H. Linkert, S. Pedersen, H. Takai, J.O. Johnsen, P.W.G. Groot Koerkamp, G.H. Uenk, J.H.M. Metz, T. Hinz, V. Caspary, and S. Linke. 1998. Emissions of aerial pollutants in livestock buildings in Northern Europe: Overview of a multinational project. Journal of Agricultural Engineering Research 70:3-9.

Watson, J.G., N.F. Robinson, C. Lewis, and T. Coulter. 1997. Chemical Mass Balance Receptor Model Version 8 (CMB8) User's Manual. Desert Research Institute, Document No. 1808.1D1 Reno, Nev.

Watson, J.G., J.C. Chow, and E.M. Fujita. 2001. Review of volatile organic compound source apportionment by chemical mass balance. Atmospheric Environment 35:1567-1584.

Watts, P.J. 1991. The Measurement of Odours: A Discussion Paper for Australia. AMLRDC Report No. DAQ. 64/5. Feedlot Services Group, Queensland Department of Primary Industries, Toowoomba. 40 pp.

Weinberg, Philip. 2000. Pollution Control. Part I of Environment Law—USA. M. R. Grossman et al. (eds.) International Encyclopedia of Laws, 2nd ed. The Hague: Kluwer Law International.

Westberg, H., and P. Zimmerman. 1993. Analytical methods used to identify nonmethane organic compounds in ambient atmospheres. Advances in Chemistry Series 232:275.

Westerman, P. W., and R. Zhang. 1995. Aerobic treatment of animal waste for odor control. Pp. 218-222 in Proceedings of International Livestock Odor Conference. Iowa State University, Ames.

Whalen, S.C., R.L. Phillips, and E.N. Fischer. 2000. Nitrous oxide emission from an agricultural field fertilized with liquid lagoonal swine effluent. Global Biogeochemical Cycles 14:545-558.

Whitman, W.G. 1923. A preliminary experimental confirmation of the two-film-theory of gas adsorption. Chemical and Metallurgical Engineering 29:146-148.

Wiebe, H.A., K.G. Anlauf, E.C. Tuazon, A.M. Winer, H.W. Biermann, B.R. Appel, P.A. Solomon, G.R. Cass, T.G. Ellestad, K.T. Knapp, E. Peake, C.W. Spicer, and D.R. Lawson. 1990. A comparison of measurements of atmospheric ammonia by filter packs, transition-flow reactors, simple and annular denuders and Fourier transform infrared spectroscopy. Atmospheric Environment 24A:1019-1028.

Wilkerson, V.A., D.P. Casper, D.R. Mertens, and H.F. Tyrrell. 1994. Evaluation of several methane prediction equations for dairy cows. In Proceedings of the 13th Symposium on Energy Metabolism of Farm Animals, J. F. Aguilera (ed.). EAAP Publ. 76. Mojacar, Spain.

Williams, A.G. 1984. Indicators of piggery slurry odor offensiveness. Agricultural Wastes 10:15-36.

Williams, C.M., and S.S. Schiffman. 1996. Effect of liquid swine manure additives on odor parameters. Pp. 409-412 in International Conference on Air Pollution from Agricultural Operations. Kansas City, Mo.: American Society of Agricultural Engineers.

Williams, C.M. 2001. Environmentally Superior Waste Management Technologies. In the Proceedings of the Agricultural Outlook Forum.

Williams, D.L., P. Ineson, and P.A. Coward. 1999. Temporal variations in nitrous oxide fluxes from urine-affected grassland. Soil Biology and Biochemistry 31:779-788.

Williams, E.J., A. Guenthre, and F.C. Fehsenfeld. 1992a. An inventory of nitric oxide emissions from soils in the United States. Journal of Geophysical Research 97:7511-7519.

Williams, E.J., S.T. Sandholm, J.D. Bradshaw, J.S. Schendel, A.O. Langford, P.K. Quinn, P.J. Lebel, S.A. Vay, P.D. Roberts, R.B. Norton, B.A. Watkins, M.P. Buhr, D.D. Parrish, J.G. Calvert, and F.C. Fehsenfeld. 1992b. An intercomparison of five ammonia measurement techniques. Journal of Geophysical Research-Atmospheres 97:11591-11611.

Wing, S., and S. Wolf. 2000. Intensive livestock operations, health, and quality of life among eastern North Carolina residents. Environmental Health Perspectives 108:233-238.

Wolynetz, M.S., and I.R. Sibbald. 1986. Relationships among major body components of broiler chicks. Poultry Science 65:2324-2329.

Wood, S.L., K.A. Janni, C.J. Clanton, D.R. Schmidt, L.D. Jacobson, and S. Weisberg. 2001. Odor and air emissions from animal production systems. Paper No. 014043 presented at American Society for Agricultural Engineers Annual International Meeting. Sacramento, Ca.

Yamulki, S., S.C. Jarvis, and P. Owen. 1998. Nitrous oxide emissions from excreta applied in a simulated grazing pattern. Soil Biology and Biochemistry 30:491-500.

Yienger, J.J., and H. Levy II. 1995. Empirical model of soil biogenic NO_x emissions. Journal of Geophysical Research 100:11447-11464.

Zahn, J.A., J.L. Hatfield, Y.S. Do, A.A. DiSpirito, D.A. Laird, and R.L. Pfeiffer. 1997. Characterization of volatile organic emissions and wastes from a swine production facility. Journal of Environmental Quality 26:1687-1696.

Zahn, J.A., A.E. Tung, B.A. Roberts, and J.L. Hatfield. 2001. Abatement of ammonia and hydrogen sulfide emissions from a swine lagoon using a polymer biocover. Journal of the Air and Waste Management Association 51:562-573.

Zahn, J.A., A.E. Tung, and B.A. Roberts. 2002. Continuous ammonia and hydrogen sulfide emission measurement over a period of four seasons from a central Missouri swine lagoon. In ASAE Paper No: 024080 presented at American Society of Agricultural Engineers Annual International Meeting/CIGR XVth World Congress, Chicago, Ill., July 28-31.

Zering, K. 1998. Comparing costs of environmental systems. In Extension Swine Educators and Adult Educators Conference. National Pork Producers Council, Des Moines, Iowa.

Zering, K. 1999. Economic Considerations in Manure Management. In Proceedings of Animal Residuals Management Conference. Water Environment Federation, Alexandria, Va. 15 pp.

Zhang, R., and P.W. Westerman. 1995. Solid-liquid separation for animal waste. Pp. 228-233 in Proceedings of International Livestock Odor Conference 1995. Iowa State University, Ames, Iowa.

Zhang, R.H., D.L. Day, L.L. Christianson, and W.P. Jepson. 1994. A computer model for predicting ammonia release rate from swine manure pits. Journal of Agricultural Engineering 58:223-229.

Zhu, J., L.D. Jacobson, D.R. Schmidt, and R.E. Nicolai. 1999. Daily Variation in Odor and Gas Emissions from Animal Facilities. ASAE Paper No. 99-4146. St. Joseph, Mich.: American Society of Agricultural Engineers.

Zhu, J., L. Jacobson, D. Schmidt, and R. Nicolai. 2000. Daily variations in odor and gas emissions from animal facilities. American Society of Agricultural Engineers 16:153-158.

Appendixes

A

Statement of Task

An ad hoc committee of the standing Committee on Animal Nutrition will be appointed to conduct a rigorous scientific review of air emission factors as related to current animal feeding and production systems in the United States. The committee will review and evaluate the scientific basis for estimating the emissions of various air pollutants (PM, PM10, PM2.5, hydrogen sulfide, ammonia, odor, VOCs, methane, and nitrous oxide) from confined livestock and poultry production systems to the atmosphere. In its evaluation, the committee will review characteristics of agricultural animal industries, methods for measuring and estimating emissions, and potential best management practices, including costs and technologic feasibility. The committee will focus on confined animal feeding production systems and will evaluate them in terms of biologic systems. The committee will consider all relevant literature and data, including reports compiled by the EPA and USDA on air quality research, air emissions, and air quality impacts of livestock waste. The study will identify critical short- and long-term research needs and will provide recommendations on the most promising science-based methodologic and modeling approaches for estimating and measuring emissions—including deposition, rate, cycle, fate, and transport—as well as on potential mitigation technologies. The committee will issue an interim report including a review of methodologies and data presented in "Air Emissions From Animal Feeding Operations" EPA Office of Air and Radiation, August 15, 2001.

B

Acronyms and Glossary

AALA: American Agricultural Law Association

Accuracy: The ability of a measurement to match the actual value of the quantity being measured.

AED: Aerodynamic equivalent particle diameter

AER: Allowable emission rate

AFO: Animal feeding operation. As defined by the U.S. Environmental Protection Agency (40 CFR 122.23), a "lot or facility" where animals "have been, are, or will be stabled or confined and fed or maintained for a total of 45 days or more in any 12 month period and crops, vegetation, forage growth, or post-harvest residues are not sustained in the normal growing season over any portion of the lot or facility."

Anthropogenic: Caused by humans.

APA: Administrative Procedure Act

APRP: Air Pollution Regulatory Process

ARS: Agricultural Research Service (USDA)

ASAE: American Society of Agricultural Engineers

ASM: Aerosol mass spectrometer

ASTM: American Society for Testing and Materials

Atmospheric stability: a property that depends on inversion strength—how rapidly air temperature rises with altitude (in units of degrees Celsius per 100 m). Strong inversions near the ground tend to stabilize the atmosphere, trap emissions, and result in higher pollutant concentrations. For a discussion of meteorological effects on carbon monoxide concentrations, see NRC (2002b).

AU: Animal unit: A unit of measure used to compare different animal types.

- EPA (66 Fed. Reg. 2960-3138): 1 cattle excluding mature dairy and veal cattle; 0.7 mature dairy cattle; 2.5 swine weighing more than 55 pounds; 10 swine weighing 55 pounds or less; 55 turkeys; 100 chickens; and 1 veal calf.
- USDA: 1000 pounds of live animal weight.

BACT: Best achievable control technology

bar: A unit of pressure equal to one atmosphere (14.7 pounds per square inch).

BAT: Best available technology (economically achievable)

Bioaerosol: Particulate matter in the atmosphere containing materials of biological origin that may cause disease, such as toxins, allergens, viruses, bacteria, and fungi.

BMP: Best management practice

BOD: Biochemical oxygen demand

BPT: Best practicable control technology (currently available)

BST: Bovine somatotropin

BW: Body weight

C: Carbon

C_2, C_{10}, C_{16}: Hydrocarbons with 2, 10, and 16 carbon atoms, respectively.

CAA: Clean Air Act

CAFO: Concentrated animal feeding operation (see Appendix E).

CCC: Commodity Credit Corporation, a government-owned and operated corporation, was created in 1933 to stabilize, support, and protect farm income and prices. It now operates as a federal corporation within USDA; the Secretary of Agriculture chairs its Board of Directors. CCC programs are carried out through the personnel and facilities of other USDA organizations, including the Farm Service Agency and the Natural Resources Conservation Service. CCC funds are used for a number of conservation programs, include EQIP (67 Fed. Reg. 48431), the Conservation Reserve Program, the Wetlands Reserve Program, and the Farmland Protection Program.

CCN: Cloud condensation nuclei

CERCLA: Comprehensive Environmental Response, Compensation, and Liability Act

cfm: Cubic feet per minute

CFR: Code of Federal Regulations

CFU: Colony forming units (bacteria formed on nutrient media)

CH_4: Methane

CNMP: Comprehensive Nutrient Management Plan

CO_2: Carbon dioxide

COD: Chemical oxygen demand

CRP: Conservation Reserve Program

CSP: Conservation Security Program

CSREES: Cooperative State Research, Education, and Extension Service (USDA)
CWA: Clean Water Act
CZARA: Coastal Zone Act Reauthorization Amendments
CZMA: Coastal Zone Management Act
Denitrification: Reduction of nitrates or nitrites to nitrogen-containing gases.
DHIA: Dairy Herd Improvement Association
DM: Dry matter
DMI: Dry matter intake
DNA: Deoxyribonucleic acid
DOAS: Differential optical absorption spectroscopy
dscm: Dry standard cubic meter
Electronic nose: An array of gas sensors that are combined with pattern recognition software to mimic human olfactory response (Lacey, 1998).
ELG: Effluent limitation guideline
Emission flux: The rate of mass emission per unit of area (e.g., tonnes per hour per hectare), typically from an area such as a waste lagoon or field.
Emission inventory: A list showing the sources and amounts (e.g., tonnes) of a pollutant emitted from a defined area for a period of time, usually one year.
Emission rate: The rate of mass emission (e.g., tonnes per hour).
EPA: U.S. Environmental Protection Agency
EPCRA: Emergency Planning and Community Right-to-Know Act
EQIP: Environmental Quality Incentives Program
Feedlot: An animal feeding operation where beef cattle are finished to slaughter weight; it consists of fenced earthen or concrete paddocks with cattle having little of no access to pasture.
FEP: Tetrafluoroethylene-hexafluoropropylene copolymer
FID: Flame ionization detector
FPM: Flame photometric detector
FRM: Federal reference method
FSA: Farm Service Agency (USDA)
ft^2: Square feet
FTIR: Fourier Transform Infrared Spectroscopy
g: Gram
g/cm^3: Grams per cubic centimeter
GC: Gas chromatography
GC-EC: Gas chromatography with electron capture detection
GC-FID: Gas chromatography with flame ionization detection
GC-MS: Gas chromatography coupled with mass spectrometry
GLAS: Ground-level area source
ha: Hectare; an area 100 meters square, about 2.5 acres
HAP: Hazardous air pollutant
HC: Hydrocarbon

HNO_3: Nitric acid
H_2S: Hydrogen sulfide
H_2SO_4: Sulfuric acid
Hz: Hertz (cycles per second)
IPCC: Intergovernmental Panel on Climate Change
ISCST: Industrial Source Complex Short Term
kg: Kilogram, or 1000 grams (about 2.2 pounds)
km: Kilometer, or 1000 meters
kwh: Kilowatt-hour
L: Liter
LAER: Lowest achievable emission rate
lbs: pounds
LD-50: The dose lethal to 50 percent of the laboratory animals tested.
Lidar: Light detection and ranging. A device similar to radar except that it emits pulsed laser light rather than microwaves.
LLPS: Low-level point source
LOAEL: Lowest observed adverse effect level
LU: Live unit, 500 kg of body weight
m: Meter
m^3: Cubic meter
MACT: Maximum achievable control technology
Manure: A mixture of animal feces and urine, which may also include litter or bedding materials.
MCF: Methane conversion factor
MeSH: Methanethiol
Mg: megagram. An SI unit of mass equal to 1 million grams or 1000 kg. This means that the megagram is identical to the tonne (metric ton). Large masses are almost always stated in tonnes in commercial applications, but megagrams are often used in scientific contexts. One megagram equals about 2204.623 pounds.
$\mu g/m^3$: Micrograms per cubic meter
μm: Micrometer or micron (10^{-6} meter)
MMD: Mass median diameter
MMTCE: Millions of metric tonnes carbon equivalent (used to express the greenhouse effect of methane and other gases relative to carbon dioxide).
MS: Mass spectroscopy
MUN: Milk urea nitrogen
mV: millivolts
MWPS: Midwest Plan Service (an organization of extension and research agricultural engineers).
N: Nitrogen
N_2: Molecular nitrogen
NA: Nonattainment area

NAAQS: National Ambient Air Quality Standards
$NaNO_3$: Sodium nitrate
NBS: National Bureau of Standards (now NIST)
NESHAPS: National Emission Standards for Hazardous Air Pollutants
NH_3: Ammonia
NH_4^+: Ammonium ion
NIST: National Institute of Standards and Technology
Nitrification: Oxidation of ammonia or an ammonium ion compound to nitric acid, nitrous acid, or any nitrate or nitrite, especially by the action of nitrobacteria.
NMHC: Nonmethane hydrocarbon
NO: Nitric oxide
N_2O: Nitrous oxide
NO_2: Nitrogen dioxide
NOAEL: No observed adverse effect level
NODA: Notice of data availability
NOV: Notice of violation
NO_x: NO and NO_2 (rapidly interconverted in the atmosphere)
NO_y: All oxidized nitrogen species in the atmosphere
NPDES: National Pollutant Discharge Elimination System
Nr: Reactive nitrogen (all nitrogen other than N_2). The term reactive nitrogen (Nr) is used in this report to denote all biologically active, photochemically reactive, and radiatively active nitrogen compounds in the atmosphere and biosphere of the earth and to distinguish all reactive forms of nitrogen from nonreactive gaseous dinitrogen (N_2). Thus, Nr includes (1) inorganic reduced forms of nitrogen (e.g., NH_3, NH_4^+), (2) inorganic oxidized forms of nitrogen (e.g., NO_x, HNO_3, N_2O, NO_3^-), and (3) a wide variety of organic nitrogen compounds including urea, amino acids, amines, proteins, nucleic acids, and so forth.
NRC: National Research Council
NRCS: Natural Resources Conservation Service (USDA)
NSPS: New Source Performance Standards
NSR: New source review
NTRM: NIST Traceable Reference Material
NUE: Nitrogen utilization efficiency (the ratio of nitrogen in animal product to nitrogen in feed consumed).
Nutrient excretion factor: An estimate of the amount of a nutrient element (e.g., N) excreted, usually reported as kilograms per day (or year) per animal (or animal unit or kilograms of body weight).
ODT: Odor detection threshold. The minimum concentration of odorant(s) detectable by 50 percent of the population (represented by an odor panel).
ORP: Oxidation-reduction potential

OU: Odor unit. The amount of odorant(s) in 1 m^3 of air detectable by 50 percent of the population

OU_E: European odor unit is the amount of odorant(s) that, when evaporated into 1 cubic meter of neutral gas at standard conditions, elicits a physiological response from a panel (detection threshold) equivalent to that elicited by 1 European Reference Odor Mass (erom), evaporated in 1 m^3 of neutral gas at standard conditions. One erom is equivalent to 123 mg *n*-butanol (CAS 71-36-3).

PAN: Peroxyacetyl nitrate

PBL: Planetary boundary layer

PCR: Polymerase chain reaction

PET: Polyethyleneterephalate (Nalophan)

PF-LIF: Photolytic fragmentation laser-induced fluorescence

PM: Particulate matter

PM2.5: Particulate matter having an aerodynamic diameter of 2.5 μm or less.

PM10: Particulate matter having an aerodynamic diameter of 10 μm or less.

PNP: Permit nutrient plan

Point source: "[A]ny discernible, confined and discrete conveyance, including but not limited to any pipe, ditch, channel, . . . concentrated animal feeding operation, . . . from which pollutants are or may be discharged. This term does not include agricultural stormwater discharges and return flows from irrigated agriculture" (33 USC § 1362(14)).

ppb: Parts per billion (by volume)

ppd: Pounds per day

ppm: Parts per million (by volume)

Precision: The degree of agreement between two or more results on the same property of identical test material expressed as the repeatability or reproducibility of an instrument reading the results.

PSD: Particle size distribution

PSD: Prevention of significant deterioration

psi: Pounds per square inch

PTFE: Polytetrafluoroethylene (Teflon)

PVF: Polyvinylfluoride (Tedlar)

REA: Relaxed eddy accumulation

RHA: Rolling herd average

RNA: Ribonucleic acid

ROG: Reactive organic gase

RQ: Reportable quantity

S: Sulfur gas

SAPRA: State air pollution regulatory agency

SCD: Sulfur chemiluminescence detector

SF_6: Sulfur hexafluoride (used as an atmospheric tracer)

SIP: State implementation plan
SO_2: Sulfur dioxide
SPM: Single point air monitor
SPME: Solid-phase microextraction. A method used to concentrate the components in odor samples prior to analysis.
SRM: Standard reference material
STPD: Standard temperature and pressure, dry
Synoptic: Relating to weather conditions that exist simultaneously over a large area.
TAN: Total ammoniacal nitrogen
TDLS: Tunable diode laser spectroscopy
Tg: Teragram. 1×10^{12} g, or 1 million metric tonnes
TKN: Total Kjeldahl nitrogen
TMDL: Total maximum daily load
TNRCC: Texas Natural Resource Conservation Commission
tpy: Tons (short) per year
TRS: Total reduced sulfur (includes H_2S and mercaptans)
TS: Total solids
TSP: Total suspended particulates
Uncertainty: The estimated amount or percentage by which an observed or calculated value may differ from the true value.
USC: United States Code
USDA: U.S. Department of Agriculture
USDC: U.S. Department of Commerce
VFA: Volatile fatty acid
VOC: Volatile organic compound
VS: Volatile solids. The weight lost upon ignition at 550 °C—an approximation of moisture and organic matter present (using Method 2540 E of the American Public Health Association).

C

Bioaerosols

Bioaerosols consist of biological compounds—mainly cell debris, viruses, bacteria, and fungi—suspended in the air. The cell debris and microbial organisms become aerosolized to form bioaerosols originating from animal respiration, skin, fur, feathers, and manure. Bioaerosols are often present at varying concentrations that are dependent on a number of factors, including emission rates, wind speed, and precipitation. Bioaerosols are thought to be a cause of concern for human health primarily because of the pathogenic viruses and bacteria they may contain. Bioaerosols may be inhaled into the lungs, where they can cause infections and allergic reactions. The people most at risk are farm or plant workers, who are in close proximity to animals and manure for extended periods of time. Aerosolization might occur from animal housing, manure storage, or land to which manure is applied. A recent National Research Council report on the land application of human biosolids found that there is a lack of scientific evidence that human settlements near these land application areas show adverse health effects (NRC, 2002c). That report recommends that the U.S. Environmental Protection agency "conduct studies that examine exposure and potential health risks to workers and community populations" from application of biosolids.

The most pressing issue with regard to bioaerosols is the challenge associated with their measurement and monitoring in the outdoor environment. The difficulty is due in large part to the diversity of organisms that can occur in a given quantity of air. One potential solution to this problem is the use of indicator species. Several technologies and methods exist to accurately measure aerosol emissions. Among these technologies are the aerosol mass spectrometer (ASM) and the electronic nose. Other methods used include bacterial culturing and molecular or chemical techniques.

Reducing the levels of pathogens before they become aerosolized is an option that would reduce bioaerosol pathogenicity. Several different methods can be used to accomplish this: one is to reduce the pathogen levels in the animal through vaccination, antibiotic therapy, diet modification, or on-farm hygiene and sanitation. Other processes focus on the inactivation of pathogens in animal waste. Inactivation of pathogens in manure can be achieved through its conversion to salable fertilizer, which may involve chemical disinfection and composting. The use of lagoons and biofilters may also provide an effective treatment. The treatment techniques that are most effective at reducing pathogen levels may be the most costly, so that reducing these costs will be important. Another important area for study is the impact of bioaerosols from animal operations on human health. Extensive studies have been undertaken on the impacts of bioaerosols and pathogens on human health, but these studies have focused on the indoor environment in places such as offices, hospitals, and animal sheds.

Until further research has been done, conclusions about the health issues surrounding bioaerosols cannot be reached; likewise, conclusions and recommendations concerning bioaerosol emissions cannot be formulated.

D

Nitrogen and Sulfur Contents of Animal Products and Live Animals—Sample Excretion Predictions

The process-based, mass balance approach would begin by predicting nitrogen, carbon, and sulfur in manure excreted. This prediction would be made by subtracting the quantities of these elements in animal products from the quantities consumed. For every major type of farm animal and every production group within these types, such predictions of intake are already available. Current publications from the National Research Council (NRC, 1994, 1998a, 2000, 2001a) detail nutrient requirements for various animal types and production systems and with varying amounts of production. Moreover, these publications have been updated periodically. Tables D-1 to D–3 are based on the assumption that animals are fed to meet National Research Council recommendations. These tables show that different types of animals convert feed nutrients to human-consumable products at differing efficiencies. The whole-system analysis also requires understanding that cattle, which appear to use feed nutrients least efficiently, in fact consume whole plant feeds (forages) that can be produced with lower environmental impact or by-products that might otherwise be a waste product.

Producers can maintain good records of the quantities of animal products sold and the nutrient composition determined from protein content. Thus, for any given animal feeding operation (AFO), manure nutrient output can be estimated from the number of animals of each type and their average production. Since some farms feed more or less of certain nutrients than the National Research Council recommends, a more accurate estimate of manure output can be made by quantifying the actual feed inputs and the export of animal farm products. Farm feed and export receipts can be used to document this balance, or diets formulated and feed composition can be used.

TABLE D-1 Typical Nitrogen and Sulfur Content of Animal Products

Product	N (%)	S (%)
Milk (% of milk weight)	0.5	0.023
Eggs (% of whole egg weight including shell)	1.78	0.16
Live Cattle		
At <30% of mature weight	2.9	0.19
Growing (30-80% of mature weight)	2.6	0.17
Finished cattle	2.0	0.13
Mature breeding cattle	2.2	0.15
Live Swine		
Nursing piglet	2.0	0.13
Growing (6-80% of mature weight)	2.3	0.15
Finished pig	2.0	0.13
Mature breeding pig	2.2	0.15
Live Poultry		
Starters	2.6	0.17
Growers	2.5	0.17
Finished broiler	2.3	0.15
Layers and breeders	2.4	0.16

CALCULATIONS AND ASSUMPTIONS FOR TABLE D-1

Milk

Approximately 93 percent of milk nitrogen is contained in true protein, while the remainder is found in nonprotein components. Therefore, milk crude protein can be calculated from milk true protein by dividing by 0.93 (NRC, 2001a). Milk nitrogen (N) can be calculated by dividing milk crude protein by 6.38 (USDA, 1941). Milk protein contains 2.4 g/16 g N as methionine and 0.87 g/16 g N as cystine (Hurrel et al., 1980). Methionine in a peptide is 21.8 percent sulfur (S), while cystine is 23.7 percent. Thus, the N:S ratio is 21.9 g/g, and sulfur in milk was determined by dividing the weight of nitrogen by 21.9.

Eggs

For a medium egg (mass = 58 g), the edible portion is 51.6 g (Royal Society of Chemistry, 1991), which means that 11 percent of the whole egg is shell. The edible portion is 65, 50, 44 and 37 g for jumbo, large, medium, and small eggs, respectively (USDA, 2002b), or 73, 56, 49, and 42 g for eggs including shells. The edible portion is 12.4 to 12.6 percent protein and 0.18 percent sulfur (Royal Society of Chemistry, 1991; USDA, 2002b). Nitrogen is calculated as crude protein divided by 6.25 (USDA, 1941).

Live Weight

Nitrogen is determined as protein content divided by 6.25. Sulfur is determined as nitrogen divided by 15. As animals grow, water content decreases while fat content increases. Protein initially increases due to decreasing water and then decreases due to increasing fat. On a dry, fat-free basis, protein comprises 80 percent of most animals' empty (not including gut contents) body weight. The change in protein as cattle and pigs age is shown in Figure D-1. The curve for swine was developed from the data published by Mahan and Shields (1998). The curve for cattle was derived by integrating the change in total body protein (protein accretion) and body weight gain predicted by the National Research Council (NRC, 2000) and dividing the former values by the latter. Ferrell and Jenkins (1998) reported protein percentage of live weight for a variety of breeds of beef cattle fed differently to 80 percent of mature weight to range from 12.4 to 13.8 percent. Hutcheson et al. (1997) reported protein percentage of finished Brangus steers to range from 12.6 to 13.1 percent of live weight. For mature cattle, a

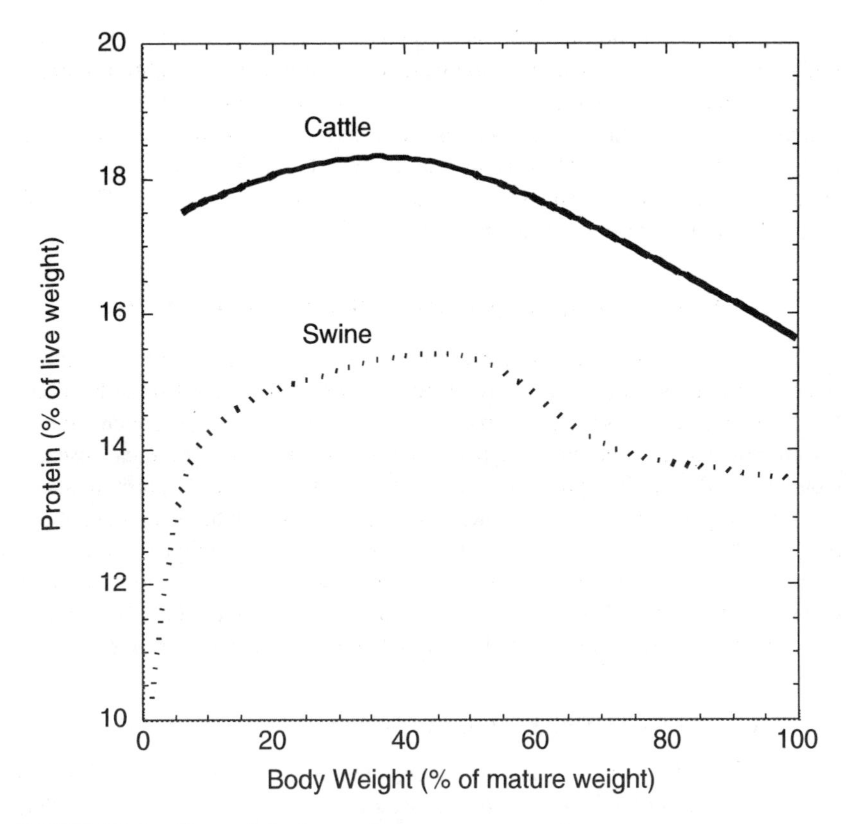

FIGURE D-1 Change in body protein percentage as cattle mature.

TABLE D-2 Nitrogen and Sulfur Content of Animal Live Weight Gain[a]

Animal Type	N (%)	S (%)
Cattle		
<30% of mature weight	3.0	0.20
40% of mature body weight	2.8	0.19
60% of mature body weight	2.4	0.16
80% of mature body weight	2.0	0.12
Swine		
<35% of mature weight	2.4	0.16
35-80% of mature weight	1.8	0.12
Poultry		
Growing broilers	2.3	0.15

[a]Grams per 100 grams of live weight gain.

model using body condition score is recommended (NRC, 2001a) for predicting body protein directly. Breeding cattle were assumed to have a condition score of 3.0 on a 5-point scale, and finished cattle were assigned a score of 4.25. Compositions of poultry carcasses as a percentage of live weight were based on reports by Wolynetz and Sibbald (1986) and Van der Hel et al. (1992) for starters, Brady et al. (1978) for growing, and Santoso et al. (1995) for mature broilers. Composition of layers was based on Katanbaf et al. (1989).

CALCULATIONS AND ASSUMPTIONS FOR TABLE D-2

For cattle, the protein accretion associated with live weight gain was summarized by previous National Research Council (NRC, 2000, 2001a) publications. For the young calf (less than 30 percent of mature body weight), nitrogen retained for growth is assumed to be 3.0 g per 100 g live weight gain (National Research Council, 2001a). Using typical growth rates for cattle (0.83 kg/d), protein accretion per kilogram live weight gain was calculated according to the model defined by the National Research Council (2000). For swine, the protein as a percentage of live weight gain (grams per 100 grams of gain) was estimated as the derivative from Figure D-1 based on carcass data. For poultry, data on total carcass composition at maturity were used and divided by the weight change during growing.

CALCULATIONS AND ASSUMPTIONS FOR TABLE D-3

All calculations were made according to recommendations in the body of this report. Current feeding recommendations were assumed in order to calculate excretion patterns in different types of livestock. These approximations may vary

TABLE D-3 Sample Excretion Predictions[a] Directly from Different Types of Food Production Animals

Animal Type	Fecal N	Urine N	Fecal S	Fecal C
Dairy Cattle				
Small Frame (e.g. Jersey)				
Lactating, RHA[b] = 3050 kg	116	100	23	1739
Lactating, RHA = 6100 kg	164	143	28	2260
Lactating, RHA = 9150 kg	212	186	33	2780
Nonlactating, mature	104	101	24	1726
Heifers, 200 kg BW[c]	44	38	9	698
Heifers, 375 kg BW	82	98	17	1253
Large Frame (e.g. Holstein)				
Lactating, RHA = 6100 kg	236	202	39	3232
Lactating, RHA = 9150 kg	277	236	43	3684
Lactating, RHA = 12,200 kg	318	269	48	4136
Non-lactating, mature	114	98	27	1917
Heifers, 200 kg BW	59	53	12	911
Heifers, 375 kg BW	111	133	23	1671
Beef				
Small Frame (500 kg; 1.3 kg/d)				
Growing (300 kg)	71	59	10	1118
Growing (350 kg)	76	57	12	1255
Growing (400 kg)	81	52	13	1391
Pregnant heifer	68	37	14	1292
Cow and calf	81	40	15	1503
Large Frame (635 kg; 1.5 kg/d)				
Growing (380 kg)	85	69	12	1342
Growing (440 kg)	90	65	14	1503
Growing (500 kg)	96	59	16	1665
Pregnant heifer	89	74	16	1547
Cow and calf	106	81	19	1838
Swine				
Growing (1-5 kg)	2.1	3.4	0.2	17
Growing (5-20 kg)	2.9	6.2	0.31	25.2
Growing (20-120 kg)	10	24.9	1.10	121.7
Bred sow	7.5	25	1.2	126
Lactating sow and piglets	30	113	5.1	360
Boar	8.3	33	1.9	134
Poultry				
White Egg Layers				
Growing (avg. for 20 wks)	0.19	0.83	0.039	3.3
Layers	0.36	1.20	0.057	6.7
Brown Egg Layers				
Growing (avg. for 20 wks)	0.19	0.82	0.032	3.4

continues

TABLE D-3 Continued

Animal Type	Fecal N	Urine N	Fecal S	Fecal C
Layers	0.44	1.55	0.063	7.4
Meat-Type Chickens				
Broilers (avg. for 7 wks)	0.46	1.51	0.008	6.4
Roasters (avg. for 9 wks)	0.70	1.91	0.019	10
Meat-type laying hens	0.47	1.82	0.13	8.7
Meat-type breeder roosters	0.29	0.80	0.10	7.6
Turkeys				
Growing males (avg. for 24 wks)	1.77	7.3	0.095	27
Growing females (avg. for 20 wks)	1.22	5.0	0.061	17
Laying hens	0.34	0.59	0.14	19
Breeder males (22 kg)	0.29	1.63	0.52	38

[a]Grams per day per animal.
[b]RHA = rolling herd average, average lactating cow's milk production per 305-day lactation.
[c]BW = body weight.

by up to 30 percent in either direction for specific animal feeding operations, depending on feeding and management practices.

Dairy

Dry matter intake and protein feeding requirements for typically managed animals were determined (NRC, 2001a). Mature body weight was assumed to be 454 kg for small breeds (e.g., Jersey) and 680 kg for large breeds (e.g., Holstein and Brown Swiss). Heifer growth rates were assumed to be 0.5 kg/d for 100-300-kg body weight and 0.6 kg/d for 300-450-kg body weight for small breeds. Growth rates were assumed to be 0.8 kg/d for large breeds. An average lactating cow was defined for each level of herd milk production. The average cow was assumed to be multiparous and 90 days in milk. Milk from small-breed cattle was assumed to be 4.5 percent fat and 3.5 percent true protein, and milk from large-breed cattle was assumed to be 3.5 percent fat and 3.0 percent true protein. Average DMI (dry mater intake) was assumed to be in accordance with National Research Council predictions (NRC, 2001a). Crude protein was assumed to be fed at 8 percent above the average cow's requirement because producers feed for a higher level of production than the average to avoid the risk of lost milk production from higher-producing cows. At 8 percent above the average cow's requirements, protein should be sufficient for the 82nd percentile cow. In addition, protein in excess of requirements is fed to account for variation in feed composition. Nitrogen intake was equal to crude protein intake divided by 6.25.

Beef

Dry matter intake and protein feeding requirements for typically managed animals were determined (NRC, 2000). Mature body weight was assumed to be 500 kg for small breeds (e.g., Angus) and 635 kg for large breeds (e.g., Simental). Growth rates were assumed to be 1.3 kg/d for small breeds and 1.5 kg/d for large breeds. Growth rates depend on diet energy and protein concentrations and would greatly affect excretion per day. Mature cows were assumed to be six months, postcalving.

Swine

Dry matter intake and protein feeding requirements for typically managed animals were determined (NRC, 1998a). Growing pigs were assumed to be gaining 320 g lean body mass per day from 20-kg body weight to harvesting. Bred sows were assumed to weigh 140 kg at breeding. Sulfur intake was assumed to be the amount needed to meet the requirements for methionine and cystine. Apparent dry matter digestibility was assumed to be 82 percent, and carbon was assumed to represent 41.5 percent of excreted dry matter.

Poultry

Dry matter intake and protein feeding requirements for typically managed animals were determined (NRC, 1994). For meat animals, the total intake was calculated by week and for the duration of feeding; then the average excretion per day over the course of the entire production time was calculated. Broilers were assumed to be harvested at seven weeks and roasters at nine weeks. Sulfur intake was assumed to be from required sulfur amino acids multiplied by the sulfur percentages of those amino acids. For nitrogen and sulfur retention estimates for turkeys, the same composition per gram of egg was assumed as for chicken eggs, but the egg size was assumed to be 96 g (Siopes, 1999), with an average of 0.8 egg per day.

E

Animal Units

U.S. ENVIRONMENTAL PROTECTION AGENCY

The United States Environmental Protection Agency (EPA) definition of animal unit appears in the regulations that govern National Pollutant Discharge Elimination System (NPDES) permits. Appendix B to 40 CFR Part 122 defines animal unit and uses the concept to determine when an animal feeding operation is a concentrated animal feeding operation.

APPENDIX B TO PART 122—CRITERIA FOR DETERMINING A CONCENTRATED ANIMAL FEEDING OPERATION (§ 122.23)

An animal feeding operation is a concentrated animal feeding operation for purposes of § 122.23 if either of the following criteria is met.

(a) More than the numbers of animals specified in any of the following categories are confined:
   (1) 1,000 slaughter and feeder cattle,
   (2) 700 mature dairy cattle (whether milked or dry cows),
   (3) 2,500 swine each weighing over 25 kilograms (approximately 55 pounds),
   (4) 500 horses,
   (5) 10,000 sheep or lambs,
   (6) 55,000 turkeys,
   (7) 100,000 laying hens or broilers (if the facility has continuous overflow watering),

(8) 30,000 laying hens or broilers (if the facility has a liquid manure system),
(9) 5,000 ducks, or
(10) 1,000 animal units; or

(b) More than the following number and types of animals are confined:
(1) 300 slaughter or feeder cattle,
(2) 200 mature dairy cattle (whether milked or dry cows),
(3) 750 swine each weighing over 25 kilograms (approximately 55 pounds),
(4) 150 horses,
(5) 3,000 sheep or lambs,
(6) 16,500 turkeys,
(7) 30,000 laying hens or broilers (if the facility has continuous overflow watering),
(8) 9,000 laying hens or broilers (if the facility has a liquid manure handling system),
(9) 1,500 ducks, or
(10) 300 animal units;

and either one of the following conditions is met:

pollutants are discharged into navigable waters through a man-made ditch, flushing system or other similar man-made device;

or pollutants are discharged directly into waters of the United States that originate outside of and pass over, across, or through the facility or otherwise come into direct contact with the animals confined in the operation.

Provided, however, that no animal feeding operation is a concentrated animal feeding operation as defined above if such animal feeding operation discharges only in the event of a 25 year, 24-hour storm event.

The term *animal unit* means a unit of measurement for any animal feeding operation calculated by adding the following numbers:

the number of slaughter and feeder cattle multiplied by 1.0, plus the number of mature dairy cattle multiplied by 1.4, plus the number of swine weighing over 25 kilograms (approximately 55 pounds) multiplied by 0.4, plus the number of sheep multiplied by 0.1, plus the number of horses multiplied by 2.0.

The term *man-made* means constructed by man and used for the purpose of transporting wastes.

U.S. DEPARTMENT OF AGRICULTURE

A definition of animal unit from United States Department of Agriculture (USDA), appears in the Commodity Credit Corporation Regulations for the Environmental Quality Incentives Program (EQIP), at 7 CFR § 1466.3:

> *Animal unit* means 1,000 pounds of live weight of any given livestock species or any combination of livestock species.

F

Ammonia Emissions from Manure Storage

The production of ammonia from animal manure via mineralization of organic nitrogen on the kth day of storage can be predicted using the following equation:

$$TM_{TAN,k} = \Gamma N_i \cdot N_{0i} \cdot NCF_{jk} \cdot WS_{ij} \cdot CAF,$$

where

$TM_{TAN,k}$ = production of total ammoniacal nitrogen (TAN) from manure storage on the farm in grams per day on the *k*th day of storage;
N_i = total nitrogen excreted by animal *i* in grams per day;
N_{0i} = maximum TAN production potential of the manure from animal *i* in grams per grams of total nitrogen;
NCF_{jk} = nitrogen conversion factor for manure storage *j*, for *k* days of storage time, which represents the extent to which the N_0 is realized (note: $0 \leq NCF \leq 1$);
WS_{ij} = fraction of waste of animal *i* handled in manure storage *j*;
CAF = climate adjustment factor for the farm, which represents the extent to which N_0 is realized under climatic conditions (e.g., temperature and rainfall) on the farm (note: $0 \leq CAF \leq 1$).

Quantitative information on the production of ammonia in animal manure is scarce in the literature. Zhang et al. (1994) developed equations for predicting the production rate of ammonia in swine manure storage as a function of time and depth in the manure. However, their study did not account for the influence of different temperatures, manure solid content, and oxygen content. More research is needed to develop accurate prediction models for quantifying the production rate of ammonia due to mineralization of organic nitrogen in different types of manure management systems.

Once ammonia (NH_3) is generated in animal housing or manure storage, its emission from the manure to the atmosphere is controlled by the aqueous chemistry of NH_3 in the manure and the convective mass-transfer mechanism at the manure surface. The pH, manure temperature, air temperature, wind velocity, and relative humidity are major factors affecting the emission process. The pH controls the partitioning of ammonia between NH_3 and NH_4^+ (ammonium ion) in the water. The emission rate of ammonia from manure on the *k*th day can be calculated using the following mass-transfer equation,

$$M_{NH_3-N,k} = 86{,}400\ K_L\ F\ [TAN]_k,$$

where M is g/m^2, 86,400 is the number of seconds in a day, K_L is the mass transfer coefficient in meters per second, *F* is the fraction factor for free ammonia in total ammonia and has a value of 0-1, and $[TAN]_k$ is the concentration of total ammoniacal nitrogen in milligrams per liter after *k* days. *F* can be determined as a function of pH and ionization constant (K_a) of ammonia in water, using the following equation:

$$F = \frac{[NH_3]}{[TAN]} = \frac{K_a 10^{pH}}{K_a 10^{pH} + 1}.$$

K_a is a function of water temperature (T_{aq}, kelvin) as shown in the following equation:

$$K_a = \frac{[NH_3][H^+]}{[NH_4^+]} = 10^{-\left(0.0897 + \frac{2729}{T_{aq}}\right)}.$$

It has been found by researchers that the K_a in wastewater has a different value from that in water (Zhang et al., 1994; Liang et al., 2002). The K_a in animal manure ($K_{a,m}$) is 25-50 percent of the K_a in water, depending on the characteristics of manure, such as its solids content.

K_L is the convective mass-transfer coefficient and [TAN] is the concentration of total ammoniacal nitrogen at the manure surface. If the stratification of NH_3 in the manure is negligible, [TAN] can be assumed to be the concentration in the bulk liquid. For any given day *k*, $[TAN]_k$ can be calculated by the concentration at the end of previous day, $[TAN]_{k-1}$, and the ammonia generated on the kth day ($TM_{TAN,k}$, grams) divided by the volume (V, cubic meters) of the liquid manure in storage, as shown below:

$$[TAN]_k = [TAN]_{k-1} + TM_{TAN,k}/V.$$

The mass-transfer coefficient K_L is a function of manure temperature, air temperature, wind velocity, and relative humidity. Various equations for K_L are available in the literature. However, most of them were developed using controlled experiments by means of convective mass-transfer chambers, and have not been well validated using field-scale experiments. More research is needed to calibrate and validate them. An example is given below is based on the two-film theory.

AMMONIA MASS-TRANSFER COEFFICIENT

The mass-transfer coefficient for ammonia as derived from the two-film theory (Whitman, 1923) is given as follows:

$$K_L = \frac{k_{L_{NH_3}} K_H k_{G_{NH_3}}}{K_H k_{G_{NH_3}} + k_{L_{NH_3}} K_H k_{G_{NH_3}}},$$

where K_L is the overall mass-transfer coefficient in meters per second, K_H is Henry's law constant (dimensionless) calculated as a function of water temperature (T_{aq}, kelvin),

$$K_H = \left[\frac{2.395\times 10^5}{T_{aq}}\right] e^{-\frac{4151}{T_{aq}}},$$

and k_{GNH_3} and k_{LNH_3} are mass-transfer coefficients (meters per second) through gaseous and liquid films, respectively, at the interface of water and air, and are related to the diffusivities (square meters per second) of ammonia and water in air (D_{airNH_3} and D_{airH_2O}), and of ammonia and oxygen in water ($D_{waterNH_3}$ and D_{waterO_2}):

$$k_{G_{NH_3}} = k_{G_{H_2O}} \left(\frac{D_{air_{NH_3}}}{D_{air_{H_2O}}}\right)^{0.67} \qquad k_{L_{NH_3}} = k_{L_{O_2}} \left(\frac{D_{water_{NH_3}}}{D_{water_{O_2}}}\right)^{0.57}$$

$$k_{G_{H_2O}} = 5.158 \times 10^{-5} + 1.954 \times 10^{-3} u_8 \qquad k_{L_{O_2}} = 1.676 \times 10^{-6} e^{0.236 u_8}$$

where u_8 is the wind velocity (meters per second) at 8 m above the water surface. The diffusivities are calculated using the following equations:

$$D_{air_{NH_3}} = \frac{10^{-7} T_a^{1.75} \left[\frac{M_{air} + M_{NH_3}}{M_{air} M_{NH_3}} \right]^{0.5}}{P \left[(\Sigma v)_{air}^{\frac{1}{3}} + (\Sigma v)_{NH_3}^{\frac{1}{3}} \right]^2}$$

$$D_{air_{H_2O}} = \frac{10^{-7} T_a^{1.75} \left[\frac{M_{air} + M_{H_2O}}{M_{air} M_{H_2O}} \right]^{0.5}}{P \left[(\Sigma v)_{air}^{\frac{1}{3}} + (\Sigma v)_{H_2O}^{\frac{1}{3}} \right]^2}$$

$$D_{water_{O_2}} = 7.28236 \times 10^{-15} \frac{T_{aq}}{e^{\left(\frac{1622}{T_{aq}} - 12.40581 \right)}}$$

$$D_{water_{NH_3}} = 6.14526 \times 10^{-15} \frac{T_{aq}}{e^{\left(\frac{1622}{T_{aq}} - 12.40581 \right)}}$$

where M is the molecular weight (grams per mole), T_a and T_{aq} are the air and water temperatures (kelvin), and v is the molecular diffusion volume (cubic centimeters per mole).

SYMBOL DEFINITIONS, UNITS

K_L	Overall mass-transfer coefficient of ammonia, cm/h
K_H	Henry's coefficient, dimensionless
$k_{L_{NH_3}}$	Mass-transfer coefficient of ammonia in liquid phase, cm/h
$k_{G_{NH_3}}$	Mass-transfer coefficient of ammonia in gas phase, cm/h
$k_{L_{O_2}}$	Mass-transfer coefficient of oxygen in liquid phase, cm/h
$k_{G_{H_2O}}$	Mass-transfer coefficient of water in gas phase, cm/h
u_8	Wind speed at 8-m height, m/s
u_z	Wind speed at an anemometer height z, m/s
z_0	Roughness height, m
P	Atmospheric pressure, atm
$D_{air_{NH_3}}$	NH_3 diffusion coefficient in air, m^2/s
$D_{air_{H_2O}}$	H_2O diffusion coefficient in air, m^2/s
$D_{water_{O_2}}$	O_2 diffusion coefficient in water, m^2/s
$D_{water_{NH_3}}$	NH_3 diffusion coefficient in water, m^2/s
M_{air}	Molecular weight of air (average), g/mol (29)
M_{NH_3}	Molecular weight of NH_3, g/mol (17)
M_{H_2O}	Molecular weight of H_2O, g/mol (18)
$(\Sigma v)_{air}$	Air molecular diffusion volume, 20.1 cm^3/mol
$(\Sigma v)_{NH_3}$	NH_3 molecular diffusion volume, 14.9 cm^3/mol
T_a	Air temperature, K, mg/l
$[TAN]_k$	Total ammoniacal concentration at the manure surface on the *k*th day of storage
T_{aq}	Water (manure) temperature, K

G

Regulatory Action Levels by Selected Atmospheric Pollutant[a,b]

Pollutant	Action Level (tpy)	Requirement	40 CFR (2001-2002)
NH_3	18	Emission release notifications under EPCRA or CERCLA	302.4 (Table 302.4)
NO_x	250	PSD review for construction of a new major source	51.166, 52.21
	100	NA review for construction of a new major source	51.165, 52.24
	100	Part 70 operating permit in attainment or unclassified area	70.2, 71.2
	50	Part 70 operating permit in serious nonattainment area	70.2, 71.2
	40	PSD or NA review for modifications to an existing major source	51.165, 51.166, 52.21, 52.24
	25	Part 70 operating permit in severe nonattainment area	70.2, 71.2
	10	Part 70 operating permit in extreme nonattainment area	70.2, 71.2
VOC	250	PSD review for construction of a new major source	51.166, 52.21
	100	NA review for construction of a new major source	51.165, 52.24
	100	Part 70 operating permit in attainment or unclassified area	70.2, 71.2
	50	Part 70 operating permit in serious nonattainment area	70.2, 71.2
	40	PSD or NA review for modifications to an existing major source	51.165, 51.166, 52.21, 52.24
	25	Part 70 operating permit in severe nonattainment area	70.2, 71.2
	10	Part 70 operating permit in extreme nonattainment area	70.2, 71.2
Any single HAP	10	Part 70 operating permit and case-by-case MACT for new source	70.2, 63.41
Combination of all HAPs	25	Part 70 operating permit and case-by-case MACT for new source	70.2, 63.41
H_2S	250	PSD review for construction of a new major source	51.166, 52.21

	100	Part 70 operating permit	70.2
	18	Emission release notifications under EPCRA or CERCLA	302.4 (Table 302.4)
	10	PSD review for modifications to an existing major source	51.166, 52.21
PM	250	PSD review for construction of a new major source	51.166, 52.21
	100	Nonattainment review for construction of a new major source	51.165, 52.24
	25	PSD review for modifications to an existing major source	51.166, 52.21
PM10	250	PSD review for construction of a new major source	51.166, 52.21
	100	NA review for construction of a new major source	51.165, 52.24
	100	Part 70 operating permit in attainment or unclassified area	70.2, 71.2
	70	Part 70 operating permit in serious nonattainment area	70.2, 71.2
	15	PSD or NA (*sic*) review for modifications to an existing major source	51.166, 52.21

[a]EPA information, with citations added by the committee.
[b]See Appendix B for definitions.

H

Regulatory Action Levels by Regulatory Requirement and Action Status[a,b]

[a]Information from EPA. Action levels for H_2S include total reduced sulfur. VOCs and NO_x are precursors to ozone formation.

[b]Abbreviations: Att. = Attainment; Const. = Construction; Mod. = Modification; Reg. = Regulatory; Req. = Requirement; Sig. = Significant; Unc. = Unclassifiable. For other definitions see Appendix B.

Reg. Req.	Pollutant	Reg. Action Levels for Att. or Unc. Areas (tpy)	Pollutant	Area Class	Reg. Action Levels for NA Areas (tpy)
Title V Operating Permit:	VOCs or NO_x	100	VOCs or NO_x	Marginal	100
	Single HAP	10		Moderate	100
	H_2S	100		Severe	25
				Extreme	10
	PM10	100	PM10	Moderate	100
				Serious	70
New Const. NSR Permits:	VOCs or NO_x	250	VOCs or NO_x	Marginal	100
	H_2S	250		Moderate	100
				Serious	50
				Severe	25
				Extreme	10
	PM/PM10	250	PM10	Moderate	100
				Serious	70
Major Source Sig. Mod. NSR Permits:	VOCs or NO_x	40	VOCs or NO_x	Marginal	40
	H_2S	10		Moderate	40
				Serious	25
				Severe	25
	PM	25		Extreme	>0
	PM10	15	PM10		15[a]
Non-Major Source Sig. Mod. NSR Permits:	VOCs or NO_x	250	VOCs or NO_x	Marginal	100
	H_2S	250		Moderate	100
				Serious	50
				Severe	25
				Extreme	10
	PM/PM10	250	PM10	Moderate	100
				Serious	70
Case-by-case MACT: Const. of New sources (112(g))	Single HAP	10			
	All HAPs combined	25			

I

Emission Factors for a Feed Mill or Grain Elevator

The primary pollutant of concern for grain elevators and feed mills is particulate matter (PM). In general, these emissions are like those shown in Figure 4-2. The entrainment of PM in air is a consequence of pneumatic conveying and ventilation systems designed to prevent grain dust explosions by reducing concentrations of grain dust. Typically, controls are installed to reduce PM emission rates. The mass emission rate (M_i) for each exhaust i is determined by its PM concentration (C_i) (measured by source sampling) and its flow rate (Q_i) as follows:

$$M_i = C_i \times Q_i,$$

where the terms in the equation have units shown in parentheses: M_i (mass per unit time), C_i (mass per unit volume), and Q_i (volume per unit time).

An emission factor (EF_i) is determined from M_i and the processing rate (PR_i):

$$EF_i = \frac{M_i}{PR_i}$$

where EF_i is a dimensionless number (mass of pollutant per mass of feed or grain processed) and PR_i has units of mass per unit time.

Source sampling for PM10 from an emission point is accomplished using pre-collectors in series with a filter. The pre-collector for PM10 sampling will typically have a fractional efficiency curve that is log normal with a 50 percent collection efficiency (cut-point) of 10 ±1 μm aerodynamic equivalent diameter and a slope ($d_{84.1}/d_{50}$) of 1.5. This relatively flat penetration curve results in significant PM10 concentration measurement errors, which will ultimately result in incorrect emission factors.

J

Public Meeting Agendas

January 7, 2002—Washington, D.C.

1:00 p.m. **Sponsor Perspective, EPA**
Randy Waite, USEPA, OAR
Renee Johnson, USEPA, OW

1:30 **Issues at the Interface of Animal Agriculture and Air Quality**
Technical Assistance Perspectives
Thomas Christensen, Director
USDA, NRCS Animal Husbandry and Clean Water Programs Division
Societal and Environmental Considerations
Dr. Joseph Rudek, Senior Scientist
Environmental Defense
Industry Approaches and Dynamics
David Townsend, Vice President of Environmental Affairs
Premium Standard Farms Research and Development

3:15-3:30 **Break**

3:30 **Comments from Participants Registered to Present**

4:15 **Input from Other Participants**

January 24, 2002—Raleigh, North Carolina

7:00 p.m. **Roundtable Discussion with "Air Emissions from Animal Feeding Operations" Report Authors (August 15, 2001 Draft. EPA Contract No. 68-D6-0011 Task Order 71.)**

John H. Martin Jr., Hall Associates
Roy V. Oommen, Eastern Research Group
John D. Crenshaw, Eastern Research Group

8:30 **Adjourn**

January 25, 2002—Raleigh, North Carolina
Swine Air Emission Measurement and Mitigation

8:00 a.m. **Introduction**
Perry Hagenstein, Chair
NRC Committee on Air Emissions from Animal Feeding Operations

8:10 **In-ground Digester with Biogas Recovery and Electricity Generation**
Dr. Leonard Bull, Associate Director Animal and Poultry Waste Center
North Carolina State University

8:30 **Measurement of Trace-Gas Emissions in Animal Production Systems**
Dr. Lowry Harper, Research Scientist
United States Department of Agriculture

8:50 **Open Path Laser Technology/Modeling to Derive Emission Factors for Swine Production Facilities**
Dr. Bruce Harris, Research Scientist
Environmental Protection Agency

9:10 **Pathogens and Air Quality Concerns**
Dr. Mark Sobsey, Professor Environmental Sciences and Engineering
University of North Carolina

9:30 **Questions**
Robert Flocchini, Vice-Chair
NRC Committee on Air Emissions from Animal Feeding Operations

9:45 **Break**

10:00 **Permeable Lagoon Cover for Odor and Ammonia Volatilization Reduction**
Dr. Leonard Bull, Associate Director Animal and Poultry Waste Center
North Carolina State University

10:20 **Odor Quantification and Environmental Concerns**
Dr. Susan Schiffman, Professor of Medical Psychology
Duke University

10:40 **Technology for Mitigating PM and Odors from Buildings**
Dr. Bob Bottcher, Professor of Biological and Agricultural Engineering
North Carolina State University

11:00 **Annual Denuder Technology**
John T. Walker, Chemist
Environmental Protection Agency

11:20 **Additional Questions**
Robert Flocchini

11:30 **Sponsor Perspective**
Sally Shaver, Division Director
Office of Air Quality Planning and Standards
Environmental Protection Agency

11:50 **General Discussion**
Perry Hagenstein

12:00 p.m. Adjourn

February 24, 2002—Denver, Colorado

Monitoring Air Emissions Through
Microclimate Meteorological Techniques

1:30 p.m. Introduction
Perry Hagenstein, Chair
NRC Committee on Air Emissions from Animal Feeding Operations

1:40 **Surface Exchange Flux Measurements Utilizing the National Center for Atmospheric Research Integrated Surface Flux Facility**
Dr. Tony Delany, Engineer IV
Atmospheric Technology Division
National Center for Atmospheric Research

2:00 **Flux Footprint Considerations for Micrometeorological Flux Measurement Techniques**
Dr. Tom Horst
Atmospheric Technology Division
National Center for Atmospheric Research

2:20 **Micrometeorological Methods for Estimating VOC and Ammonia fluxes**
Dr. Alex Guenther, Scientist II
Atmospheric Chemistry Division
National Center for Atmospheric Research

2:40 **Analysis of Single Aerosol Particles with a Mass Spectrometer**
Dr. Daniel Murphy
Aeronomy Laboratory
National Oceanic and Atmospheric Administration

3:00 **Questions and General Discussion**
Robert Flocchini, Vice-Chair
NRC Committee on Air Emissions from Animal Feeding Operations

3:15 **Break**

Air Emission Measurement and Mitigation for Beef Feedlots

3:30 p.m. **Introduction**
Perry Hagenstein, Chair

3:40 **Odor Measurement and Mitigation**
Dr. John Sweeten, Professor and Resident Director
Agricultural Research and Extension Center
Texas A&M University

4:00 **Methane Production from Livestock and Mitigation**
Dr. Don Johnson, Professor
Department of Animal Sciences
Colorado State University

4:20 **Mitigation Technology**
Dr. Bob McGregor
Water and Waste

4:40 **Questions and General Discussion**
Robert Flocchini, Vice-Chair

5:00 **Comments from Participants Registered to Present**

5:30 **Input from Other Participants**

June 4, 2002—Sacramento, California

1:00 p.m. **Introduction**
Perry Hagenstein, Chair
Committee on Air Emissions from Animal Feeding Operations

1:05 **EPA and USDA Collaboration on Animal Feeding Operation Air Quality Research and Policy**
Tom Christensen, Director
Animal Husbandry and Clean Water Programs Division
USDA, National Resource Conservation Service

Gary Margheim, Natural Resource Manager
Strategic Natural Resources Issues

USDA, National Resources Conservation Service

2:00 **Sponsor Perspectives**

Randy Waite
Air Quality Planning and Standards Division
USEPA, Office of Air and Radiation

Tom Christensen, Director
Animal Husbandry and Clean Water Programs Division
USDA, National Resource Conservation Service

Ray Knighton, National Program Leader for Air, Soil, and Water Natural Resources and Environment
USDA, Cooperative State Research, Education, and Extension Service

2:45 **Break**

Air Emission Measurement, Mitigation, and Policy for Animal Feeding Operations in California

3:00 p.m. **Introduction**

Robert Flocchini
Vice-Chair Committee on Air Emissions from Animal Feeding Operations

3:05 **Air Emissions and Poultry Production in California**

Dr. Ralph Ernst
Department of Animal Science
University of California, Davis

3:35 **Air Emissions and Dairy Production in California**

Dr. Deanne Meyer
Department of Animal Science
University of California, Davis

4:05 **Changes in the California Agricultural Air Exemption**

Brent Newell
Center on Race, Poverty and the Environment
California Rural Legal Assistance Foundation

4:35 **California Air Resources Board**

Michael Fitzgibbon
Air Resources Board
California Environmental Protection Agency

5:05 **On-Farm Assessment and Environmental Review**

Ellen Hankes, Marketing Director
On-Farm Assessment and Environmental Review
Environmental Management Solutions, LLC

5:35 **General Discussion**
Perry Hagenstein
5:45 **Public Comments**

August 19, 2002—Washington D.C.

2:00 p.m. **Sponsor Perspectives**
Randy Waite
Air Quality Planning and Standards Division
USEPA, Office of Air and Radiation

C. Richard Amerman, National Program Leader
Natural Resources and Sustainable Agricultural Systems
USDA, Agricultural Research Service

Tom Christensen, Director
Animal Husbandry and Clean Water Programs Division
USDA, National Resource Conservation Service

K

Geographic Distribution of Livestock and Poultry Production in the United States for 1997

FIGURE K-1 Distribution of milk cows in 1997.
SOURCE: USDA (1997a).

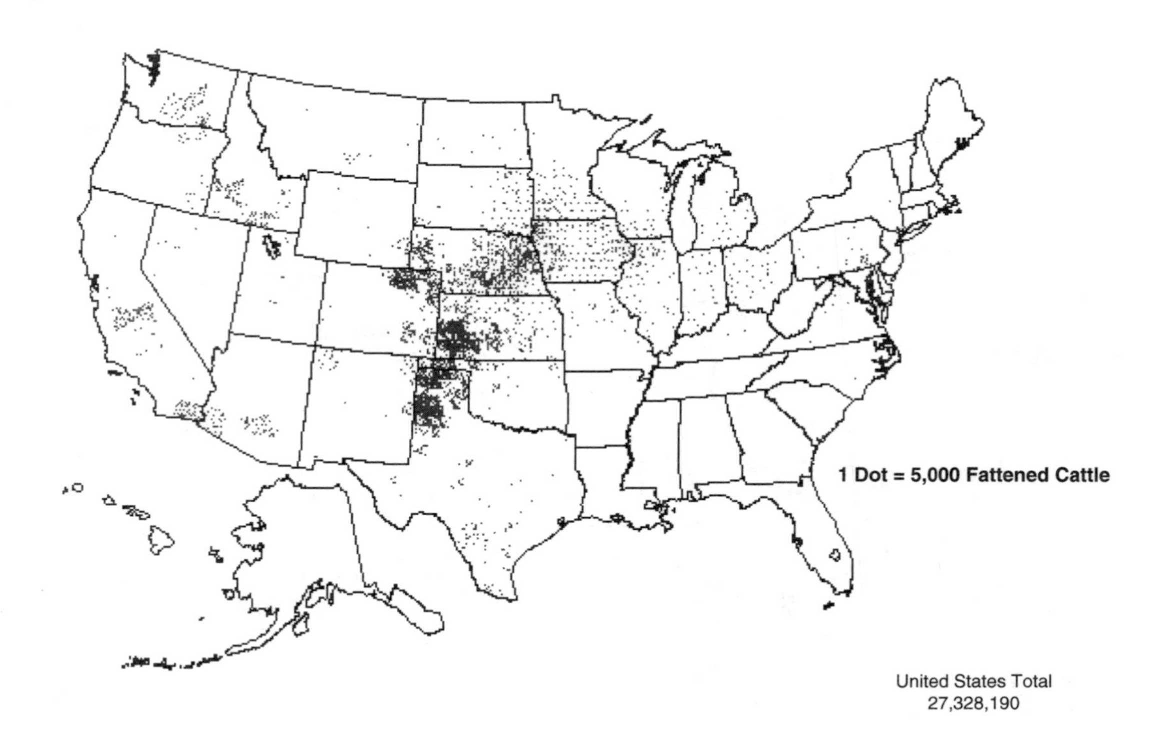

FIGURE K-2 Distribution of cattle fattened on grain and concentrates and sold in 1997.
SOURCE: USDA (1997a).

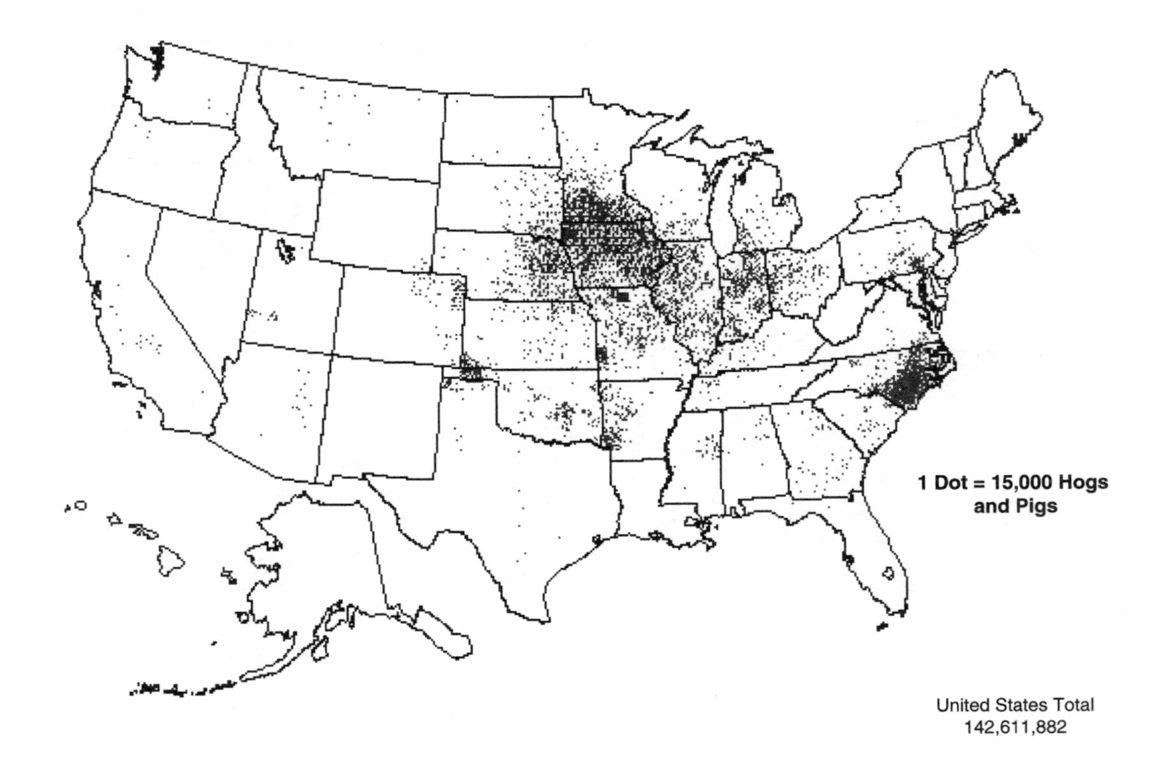

FIGURE K-3 Distribution of hogs and pigs sold in 1997.
SOURCE: USDA (1997a).

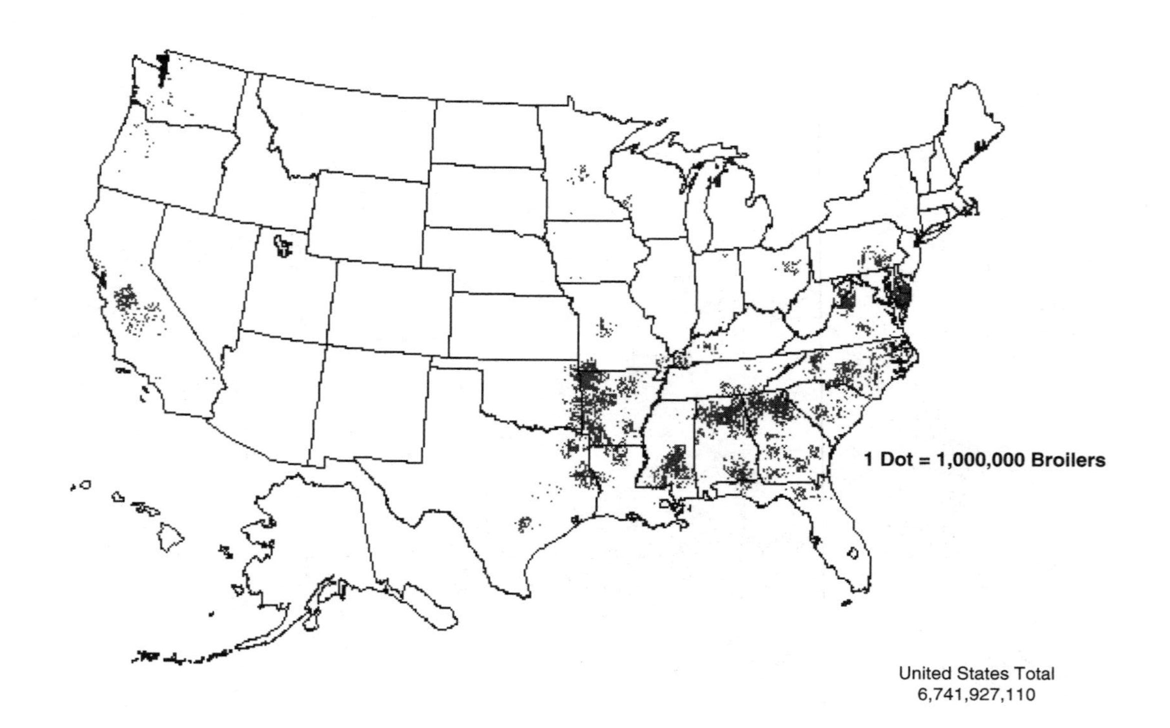

FIGURE K-4 Distribution of broilers and other meat-type chickens sold in 1997.
SOURCE: USDA (1997a).

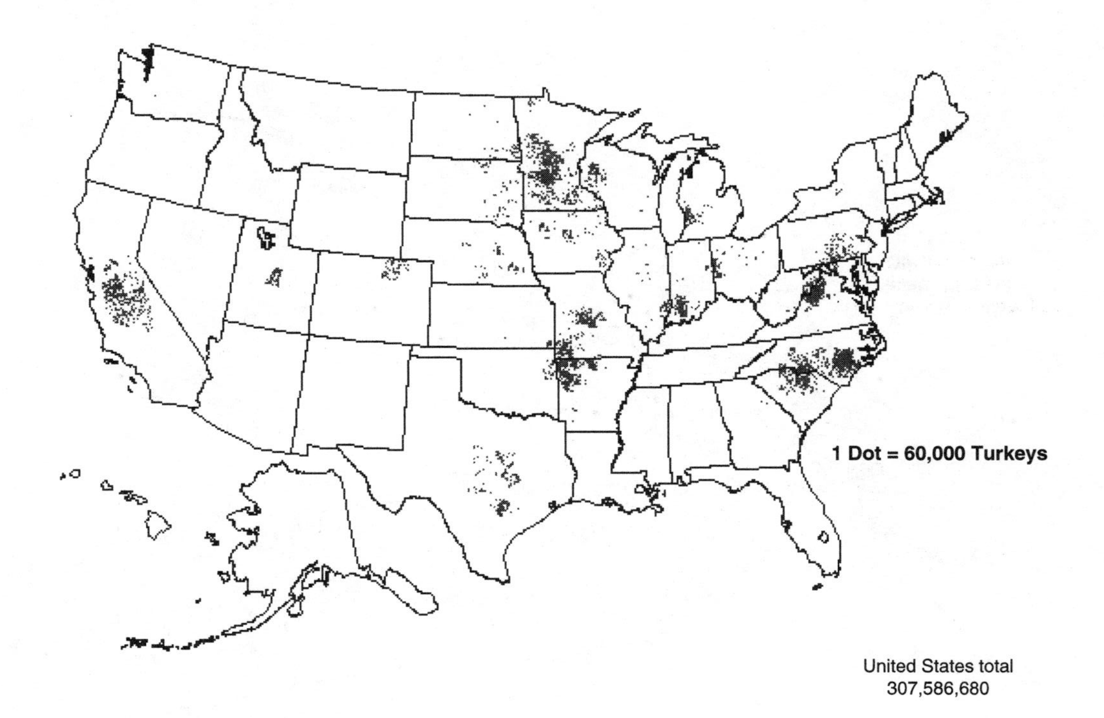

FIGURE K-5 Distribution of turkeys sold in 1997.
SOURCE: USDA (1997a).

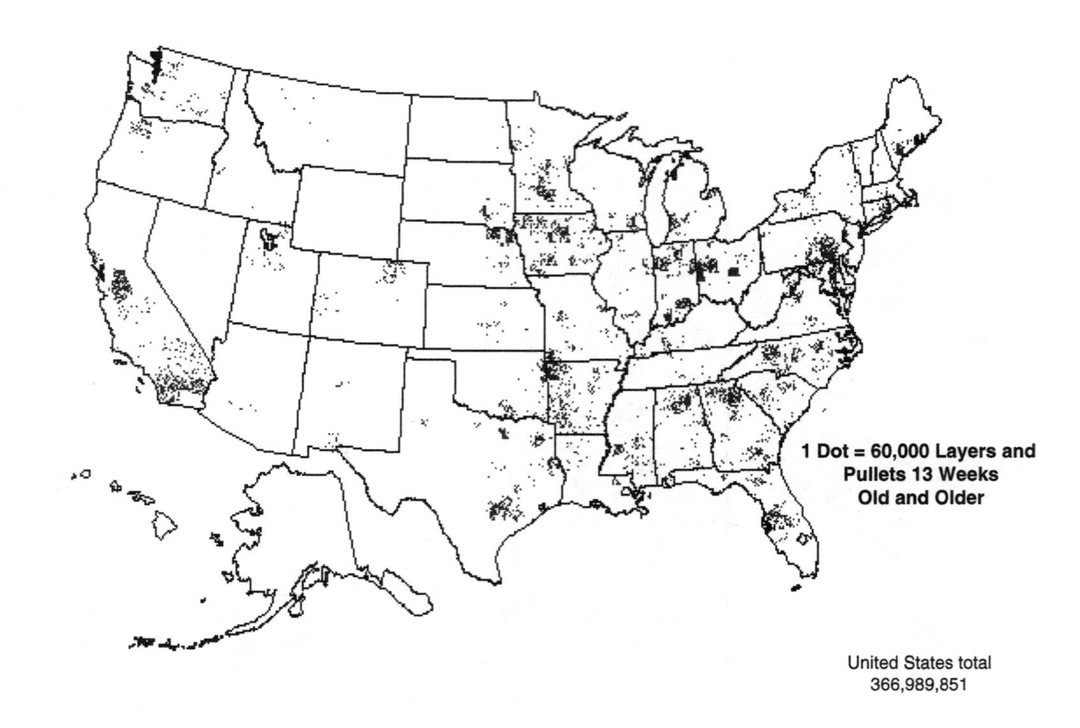

FIGURE K-6 Distribution of layers and pullets, 13 weeks old and older in 1997.
SOURCE: USDA (1997a).

L

Emission Factors in Published Literature

The following sections are excerpts from the committee's interim report *The Scientific Basis for Estimating Air Emissions from Animal Feeding Operations* (NRC, 2002a). These sections have been copy edited since the publication of the interim report.

Ammonia

Several well-designed research studies have been published establishing some of the factors that contribute to variations in ammonia (NH_3) emissions. For example, Groot Koerkamp et al. (1998) reported wide variations in emissions for different species (cattle, sows, and poultry) measured in different European countries, across facilities within a country, and between summer and fall. Amon et al. (1997) demonstrated that emissions increase as animals age. Differences due to the manure storage system have been demonstrated (Hoeksma et al. 1982). Climate, including temperature and moisture, also affects NH_3 emissions (Hutchinson et al., 1982; Aneja et al., 2000). Zhu et al. (2000) reported diurnal variation in emission measurements. With so many sources of variation in NH_3 emissions, it is unreasonable to apply a factor determined in one system, over a short period of time, to all animal feeding operations (AFOs) within a broad classification.

Although NH_3 emissions have been reported under different conditions, there are few reliable data to estimate total NH_3 emissions from all AFO components for all seasons of the year. Twenty-seven articles were used for NH_3 emission factors by EPA (2001a); of these, only eleven with original measurements were from peer-reviewed sources. Additional data were taken from six progress reports from contract research. Two of these (Kroodsma et al., 1988; North Carolina Department of Environmental and Natural Resources, 1999) were identified as "preliminary," and in one case (Kroodsma et al., 1988), the airflow measurement equipment was not calibrated.

Emission factors for NH_3 were also taken from nine review articles (EPA, 2001a); three of these modeled or interpreted previously reported information with the objective of determining emission factors (Battye et al., 1994; Grelinger, 1998; Grelinger and Page, 1999). Several of the reviews reported factors used in other countries, but not the original research used to develop them. Other reviews summarized data from primary sources that had already been considered. Thus, the review articles may not provide new information.

Most measurements and estimates reported did not represent a full life cycle of animal production. As animals grow or change physiological state, their nutrient excretion patterns vary, altering the NH_3 volatilization patterns (Amon et al., 1997). A single measurement over a short period of time will not capture the total emission for the entire life cycle of the animal. In addition, most measurements for manure storage represent only part of the storage period. The emissions from storage vary depending on length of storage, changing input from the animal system, and seasonal effects such as wind, precipitation (Hutchinson et al., 1982), and temperature (Andersson, 1998). Only one article reported measurements over an entire year (Aneja et al., 2000), although the measurements may not have been continuous. In this case, NH_3 emissions were measured from an anaerobic lagoon using dynamic flow-through chambers during four seasons. Summer emissions were 13 times greater than those in winter, and the total for the year was 2.2 kg NH_3-N (nitrogen) per animal (mean live weight = 68 kg) per year.

Expressing NH_3 emission factors on a per annum and per animal unit (AU) basis facilitates calculation of total air emissions and accounts for variation due to size of AFOs, but it does not account for some of the largest sources of variation in emissions. Clearly, there is a great deal of variation in reported measurements among AFOs represented by a single model. For example, only two references were provided for beef drylot NH_3 emission factors, but the values reported were 4.4 and 18.8 kg N/yr per animal (see EPA, 2001a, Table 8-11). For swine operations with pit storage, mean values reported in eight studies ranged from 0.03 to 2.0 kg/yr per pig of less than 25-kg body weight (see EPA 2001a, Table 8-17). This higher rate represents 66 percent of the nitrogen estimated to be excreted by feeder pigs per year (see EPA, 2001a, Table 8-10). The actual variation among AFOs represented by a single model cannot be determined without data representing the entire population of AFOs to be modeled. This would require greater replication and geographic diversity. Much of the variation among studies within a single type of model farm can be attributed to different geographic locations or seasons and the different methods and time frames used to measure the emission factors.

The approach in EPA (2001a) was to average all reported values in selected publications—both refereed and nonrefereed—giving equal weight to each article. Emission factors reported in some studies represented a single 24-hour sample, while in others, means of several samples were used. Emission factors from review articles were averaged along with the others. Properly using available data to determine emission factors, if it could be done, would require considering the uniqueness and quality of the data in each study for the intended purpose and weighting it appropriately. The causes of the discrepancies among studies would also have to be investigated.

Adding emissions from housing, manure storage, and field application, or using emission factors determined without considering the interactions of these subsystems, can easily provide faulty estimates of total emissions of NH_3. If emissions from a subsystem are increased, those from other subsystems must be decreased. For example, most of the excreted nitrogen is emitted from housing, much of the most readily available nitrogen will not be transferred to manure storage. If emissions occur in storage, there will be less nitrogen for land application. The current approach ignores these mass balance considerations and simply adds the emissions using emission factors determined separately for each subsystem.

Dividing the total manure nitrogen that leaves the farm by the total nitrogen excreted can identify some potential overestimation of emission factors. For example, using emission factors in Table 8-21 of EPA (2001a) for swine model farms, the total ammonia nitrogen emissions for 500 AUs in Model S2 can be estimated to be 1.12×10^4 kg/yr. (Three significant digits are carried for numerical accuracy from the original reference and may not be representative of the precision of the data.) The total nitrogen excreted by 500 AUs of growing hogs is 1.27×10^4 kg/yr (EPA, 2001a). Thus, one calculates that 90 percent of estimated manure nitrogen is volatilized to ammonia, leaving only 10 percent to be accumulated in sludge, applied to crops, and released as other forms of nitrogen (NO [nitric oxide], N_2O [nitrous oxide], and molecular nitrogen [N_2]). These emission factors suggest that almost all excreted nitrogen is lost as NH_3, which seems unlikely.

NITRIC OXIDE

Although nitric oxide was not specifically mentioned in the request from the U.S. Environmental Protection Agency (EPA), the committee believes that it should be included in this report because of its close relationship to ammonia. An appreciable fraction of manure nitrogen is converted to NO by microbial action in soils and released into the atmosphere. NO participates in a number of processes important to human health and the environment. The rate of emission has been widely studied but is highly variable, and emissions estimates are uncertain.

Attempts to quantify emissions of NO_x from fertilized fields show great variability. Emissions can be estimated from the fraction of the applied fertilizer nitrogen emitted as NO_x, but the flux varies strongly with land use and temperature.

Vegetation cover greatly decreases NO_x emissions (Civerolo and Dickerson, 1998); undisturbed areas such as grasslands tend to have low emission rates, while croplands can have high rates. The release rate increases rapidly with soil temperature—emissions at 30°C are roughly twice emissions at 20°C.

The fraction of applied nitrogen lost as NO emissions depends on the form of fertilizer. For example, Slemr and Seiler (1984) showed a range from 0.1 percent for $NaNO_3$ (sodium nitrate) to 5.4 percent for urea. Paul and Beauchamp (1993) measured 0.026 to 0.85 percent loss in the first six days from manure nitrogen. Estimated globally averaged fractional applied nitrogen loss as NO varies from 0.3 percent (Skiba et al., 1997) to 2.5 percent (Yienger and Levy, 1995). For the United States, where 5 Tg of manure nitrogen is produced annually, NO_x emissions directly from manure applied to soil are roughly 1 percent, or 0.05 Tg/yr, of emissions from crops used as animal feed are neglected. Williams et al. (1992a) developed a simplified model of emissions based on fertilizer application and soil temperature. They estimated that soils accounted for a total of 0.3 Tg, or 6 percent of all U.S. NO_x emissions for 1980.

Natural variability of emissions dominates the uncertainty in the estimates. In order of increasing importance, errors in land use data are about 10-20 percent, and experimental uncertainty in direct NO flux measurements is estimated at about ±30 percent. The contribution of soil temperature to uncertainty in emissions estimates stems from uncertainty in inferring soil temperature from air temperature and from variability in soil moisture. Williams et al. (1992) show that their algorithm can reproduce the observations to within 50 percent. A review of existing literature indicates that agricultural practices (such as the fraction of manure applied as fertilizer, application rates used, and tillage) introduce variability in NO emissions of about a factor of two. Variability of biomes to which manure is applied (such as short grass versus tallgrass prairie) accounts for an additional factor of three (Williams et al., 1992a; Yienger and Levy, 1995; Davidson and Klingerlee, 1997). Future research may have to focus on determining the variability of emissions, measured as a fraction of the applied manure nitrogen, with agricultural practices, type of vegetative cover, and meteorological conditions.

HYDROGEN SULFIDE

Most of the studies on hydrogen sulfide (H_2S) emissions from livestock facilities were conducted recently and included current animal housing and manure management practices. Several recent publications from Purdue University document H_2S emissions from mechanically ventilated swine buildings (Ni et al., 2002a, 2002b, 2002c, 2002d). A pulsed fluorescence SO_2 (sulfur dioxide) analyzer with an H_2S converter was used to measure H_2S concentrations in the air, and a high-frequency (16 or 24 sampling cycles each day) measurement protocol was used for continuous monitoring. In one of the studies reported, H_2S emission from two 1000-head finishing swine buildings with under-floor manure pits in Illinois was monitored continuously for a six-month period from March to September 1997. Mean H_2S emission was determined to be 0.59 kg/d, or 6.3 g/d per 500-kg animal weight. Based on emission data analysis and field observation, researchers noticed that different gases had different gas release mechanisms. Release of H_2S from the stored manure, similar to carbon dioxide and sulfur dioxide, was through both convective mass transfer and bubble release mechanisms. In comparison, the emission of NH_3 was controlled mainly by convective mass transfer. Bubble release is an especially important mechanism controlling H_2S emission from stirred manure. The differences in release mechanisms for different gases are caused mainly by differences in solubility and gas production rates in the manure. Some measurements from swine buildings were also conducted in Minnesota (Jacobson, 1999; Wood et al., 2001).

Very few data are available on H_2S emission from other types of livestock facilities such as dairy, cattle, and poultry. Using emission data from swine operations to estimate emission factors for other species such as dairy and poultry is not scientifically sound. Outside manure storage, such as storage in tanks or anaerobic lagoons, can be an important source of H_2S emissions. Emission data for such sources are lacking in the literature.

EPA (2001a) stated that H_2S emissions from solid manure systems—such as beef and veal feedlots, manure stockpiles, and broiler and turkey buildings—were insignificant, based on the assumption that these systems are mostly aerobic. Such an assumption is not valid because it is not based on scientific information. Published data indicate that a significant amount of H_2S is emitted from the composting of poultry manure when the forced aeration rate is low (Schmidt, 2000). It is very likely that H_2S is emitted from other solid manure sources as well. H_2S is produced biologically whenever there are sulfur compounds, anaerobic conditions, and sufficient moisture. Wet conditions occur in animal feedlots and uncovered solid manure piles during precipitation or in rainy seasons. Scientific studies should be conducted to provide emission data.

NITROUS OXIDE

Nitrous oxide is both a greenhouse gas and the main source of stratospheric NO_x, the principal sink for stratospheric ozone; predominantly biological processes (nitrification and denitrification) produce N_2O in soils; fertilization increases emissions. Although EPA (2001a) states that "emission factors for N_2O were not found in the literature," a large body of research exists on N_2O emissions from livestock, manure, and soils. Time constraints prevent a thorough review of the literature, but this section condenses the main points of a few recent papers and attempts to summarize the state of the science.

N_2O emissions were reviewed for the Intergovernmental Panel on Climate Change (IPCC, 2001; see also Mosier et al., 1998) with the objective of balancing the global atmospheric N_2O budget and predicting future concentrations. Although substantial uncertainties exist regarding the source strength for N_2O, agricultural activities and animal production are the primary anthropogenic sources. According to the IPCC (2001), these biological sources can be broken down into direct soil emissions, manure management systems, and indirect emissions. These three sources are about equally strong, each contributing about 2.1 Tg N/yr to the atmospheric N_2O burden. Total anthropogenic sources are estimated to be 8.1 Tg N/yr, and natural sources about 9.9 Tg N/yr, for a total of 18 Tg N/yr (Prather et al., 2001).

Soils

The Intergovernmental Panel on Climate Change estimated soil N_2O emissions as a fraction of applied nitrogen. IPCC assumed that 1.25 percent of all fertilizer nitrogen is released from soils as N_2O, with a range of 0.25 to 2.25 percent. Estimating direct soil N_2O emissions is subject to the same uncertainties as NO emissions. The fraction of applied nitrogen emitted as N_2O varies with land use, chemical composition of the fertilizer, soil moisture, temperature, and organic content of the soil. Of the global value of 2.1 Tg N/yr emitted directly from soils, Mosier et al. (1998), using the IPCC1 method, estimates that manure fertilizer contributes 0.63 Tg/yr. Using the Intergovernmental Panel on Climate Change method, 5 Tg/yr of manure nitrogen in the United States would yield 0.06 Tg N/yr as N_2O. Li et al. (1996) employed a model that accounts for soil properties and farming practices and concluded that the Intergovernmental Panel on Climate Change method underestimates emissions. They put annual N_2O emissions from all crop- and pastureland (including emissions from manure and biosolids applied as fertilizer) in the United States in the range of 0.9 to 1.1 Tg N/yr, although this number includes what Mosier et al. (1998) refers to as "indirect" sources.

Nitrification is primarily responsible for NO production, but both nitrification and denitrification lead to N_2O release from soils, and both aerobic and anaerobic soils emit N_2O. The following studies show some of the variability in estimates of the efficiency of conversion of manure nitrogen to N_2O emission. Paul and Beauchamp (1993) measured 0.025 to 0.85 percent of manure nitrogen applied to soil in the lab lost as N_2O, but Wagner-Riddle et al. (1997) found 3.8 to 4.9 percent from a fallow field. Petersen (1999) observed 0.14 to 0.64 percent emission from a barley field. Lessard et al. (1996) measured 1 percent emission of manure nitrogen applied to corn in Canada. Yamulki et al. (1998) measured emissions from grassland in England and found 0.53 percent of fecal nitrogen and 1.0 percent of urine nitrogen lost as N_2O over the first 100 days. Whalen et al. (2000) applied swine lagoon effluent to a spray field in North Carolina and observed 1.4 percent emission of applied nitrogen as N_2O. Flessa et al. (1995) applied a mixture of urea and NH_4NO_3 (ammonium nitrate) to a sunflower field in southern Germany and measured an N_2O emission of >1.8 percent of the nitrogen applied. Long-term manure application (possibly linked to increased organic content of soils) appears to increase N_2O production. Rochette et al. (2000) determined that after 19 years of manure application, 1.65 percent of applied nitrogen was converted to N_2O. Chang et al. (1998) followed the same soil for 21 years of manure application and found 2-4 percent of manure nitrogen converted to N_2O. Flessa et al. (1996) determined a total emission of N_2O from cattle droppings on a pasture equivalent to 3.2 percent of the nitrogen excreted. Clayton et al. (1994) showed that grassland used for cattle grazing could convert a larger portion of fertilizer NH_4NO_3 nitrogen to N_2O (5.1 percent versus 1.7 percent for ungrazed grassland). Williams et al. (1999) applied cow urine to pasture soil in the lab and observed a 7 percent partition of the nitrogen to N_2O.

Manure Management

Several recent studies indicate that N_2O emissions from manure can be large (Jarvis and Pain, 1994; Bouwman, 1996; Mosier et al., 1996; IPCC, 2001). For example, Jungbluth et al. (2001) measured 1.6 g N_2O/d per 500 kg of livestock emitted directly from dairy cattle; Amon et al. (2001) measured 0.62 g N_2O/d per 500 kg of livestock. Groenestein and VanFaassen (1996) found 4.8 to 7.2 g N/d per pig as N_2O.

The Intergovernmental Panel on Climate Change (2001) estimates N_2O emissions from animal production (including grazing animals) as approximately 2.1 Tg N/yr. These estimates are based on an assumed average fraction of manure nitrogen converted to N_2O and are subject to variability due to temperature, moisture content, and other environmental factors in a manner similar to soil emissions. Berges and Crutzen (1996) estimated the rate of N_2O emissions by measuring the ratio of N_2O to NH_3. They determined that 40 Tg N/yr of cattle and swine manure in housing and storage systems generates 0.2-2.5 Tg N/yr as N_2O; they did not account for additional emissions outside the housing and storage systems.

Indirect Emissions

Formation of N_2O results indirectly from the release of NH_3 to the atmosphere and its subsequent deposition as NH_3-NH_4^+ or nitrate, or from their leaching and runoff (IPCC, 2001). Human waste in sewage systems is another indirect path to atmospheric N_2O. On a global scale, leaching and runoff give an estimated 1.4 Tg N/yr; atmospheric deposition, 0.36 Tg N/yr; and human sewage, about 0.2 Tg N/yr—for a total of about 2 Tg N/yr. Dentener and Crutzen (1994) pointed out that atmospheric reactions involving NH_3 and NO_2 (nitrogen dioxide) could lead to production of N_2O; however the strength of this source is unknown.

Summary

The uncertainty in emissions of N_2O from AFOs is similar to that for NO—roughly a factor of three. While no-till agriculture decreases emissions of most greenhouse gases (Civerolo and Dickerson, 1998; Robertson et al., 2000), it appears to increase N_2O. The means for decreasing emissions do exist. Smith et al. (1997) suggested that substantial reductions in N_2O could be achieved through matching fertilizer type to environmental conditions and by using controlled-release fertilizers and nitrification inhibitors. Timing and placement of fertilizer and controlling soil conditions could also help decrease N_2O production. The vast body of work on emissions of N_2O from agricultural activities cannot be thoroughly reviewed in the short time frame of this study.

METHANE

Four original research articles, an agency report, one doctoral thesis, and one review article are cited in EPA (2001a) in estimating emission factors for methane (CH_4). Much research was overlooked since a number of papers and reports describing CH_4 emission rates can be found in the literature. Fleesa et al. (1995) reported CH_4 fluxes of 348 to 395 g per hectare (ha) per year in fields fertilized with manure. A value of 1 kg/m^2 per year CH_4 (carbon equivalents) has been reported for an uncovered dairy yard (Ellis et al., 2001). Amon et al. (2001) concluded that methane emissions were higher for anaerobically treated dairy manure than for composted manure.

EPA (2001a) estimates the CH_4 production potential of manure as the maximum quantity of CH_4 that can be produced per kilogram of volatile solids in the manure. However, a considerable amount of CH_4 is lost during eructation (belching), which this estimate does not take into account.

In estimating the CH_4 emission factor for the model farm, EPA (2001a) did not take several factors into consideration, such as the difficulty associated with measuring emissions without having a negative impact on animals. New methods have been designed to measure CH_4 emissions under pasture conditions with minimal disturbance of the animals (Leuning et al., 1999). There are some limitations to this technique; it does not work well with low wind speeds or rapid changes in wind direction, and requires high-precision gas sensors. Methane production increases while cattle are ruminating (digesting) feedstuffs—both grass and high-energy rations. In one study, lactating beef cattle grazing on grass pasture were observed to have 9.5 percent of the gross energy intake converted to CH_4 (McCaughey et al., 1999). During periods when the cattle are fed a high-grain diet, approximately 3 percent of gross intake energy is converted to CH_4 (Johnson et al., 2000).

Methods for estimating CH_4 emissions from other sources—such as rice paddies, wetlands, and tundra in Alaska—have been well studied. However, the models used to extrapolate emissions over these large areas may not apply to AFOs because of the different variables that must be taken into account. This is a knowledge gap that has to be addressed.

PARTICULATE MATTER

A limited number of studies have reported emission factors for particulate matter (PM) for various confinement systems. One of the most recent reports includes the results of an extensive study that examined PM emissions from various confinement house types, for swine, poultry, and dairy in several countries in Northern Europe (Takai et al., 1998), and a few studies report cattle or dairy drylot emissions in the United States (Parnell et al., 1994; Grelinger, 1998; Hinz and Linke, 1998; USDA, 2000b). Some of this work was cited by EPA (2001a). Two PM10 emission factors for cattle were reported for drylot feed yards by Grelinger (1998) and USDA (2000b). Another emission factor for poultry broiler house emissions was also included (Grub et al., 1965).

According to the EPA (1995b) AP-42 document, emission factor data are considered to be of good quality when the test methodology is sound, the sources tested are representative, a reasonable number of facilities are tested, and the results are presented in enough detail to permit validation. Whenever possible, it is desirable to obtain data directly from an original report or article, rather than from a compilation or literature summary. Only a very limited number of published papers have been used to estimate PM emission factors for AFOs. Some of the papers utilized do not appear to be of the highest quality or relevance to modern operations. Takai et al. (1998) and Grub et al. (1965) appeared in the peer-reviewed literature, but other work cited was not. Takai et al. (1998) represents one of the most extensive studies conducted on livestock houses to date; it made 231 field measurements of dust concentrations and dust emissions from livestock buildings across Northern Europe. Factors included in their study design were country (England, the Netherlands, Denmark, and Germany); housing (six cattle housing types, five swine housing types, and three poultry housing types); season (summer and winter); and diurnal period (day and night). Each field measurement was for a 12-hour period, and each house was sampled for a 24-hour period, or two 12-hour samples per house. Where possible, measurements were repeated at the same house for both seasons (Wathes et al., 1998).

One reference (Grelinger, 1998) appeared in a specialty conference proceedings (non-peer reviewed), and it is not clear how the emission rates were derived. The U.S. Department of Agriculture (USDA, 2000b) summarizes results from other cattle studies. The Grub et al. (1965) study was more than 35 years old and reported emission factors for a poultry confinement configuration (chambers 2.4 m by 3.0 m by 22.1 m high, ventilated at a constant airflow rate) that is not used in current operations.

The sizes of ambient particulate matter varied from study to study, ranging from "respirable" and "inhalable" to total suspended particulates (TSPs). Takai et al. (1998) sampled inhalable dust using European Institute of Occupational Medicine dust samplers. The respirable fraction was measured using cyclone dust samplers with a 50 percent cut diameter of 5 μm. Grub et al. (1965) measured dust rather than PM10; it is not clear whether the emission factors quoted represented dust or PM10 estimated from the dust. Grelinger (1998) measured TSP and obtained PM10 by multiplying by 0.25. USDA (2000b) reported that TSP was measured rather than PM10, according to the AFO project data summary sheets in EPA (2001a). The representativeness of emission factors in the literature is also questionable. For example, the emission factors reported by Takai et al. (1998) were based on data collected for very brief periods, one to two days at each barn. Relevant work was overlooked in the estimation of cattle feedlot PM emissions (e.g., Parnell et al., 1994), or it is not clear from EPA (2001a) whether that work was included in the USDA (2000b) publication cited. Auvermann et al. (2001) extensively reviewed the PM emission factors suggested for AFOs (for both feedlots and feed mills) in AP-42 (EPA, 1995b). They pointed out that the PM10 emission factor for cattle feedlots specified in AP-42 was five times as high as the more recent values determined by Parnell et al. (1994). EPA (2001a) did not discuss the AP-42 emission factors.

When more than one study was found that examined PM emissions, the results were not consistent among studies. The two poultry house emission factors differed by an order of magnitude and were simply averaged to characterize PM emissions from poultry houses, even though the Grub et al. (1965) study was of questionable relevance to today's production systems. The two drylot cattle yard PM emission factors differed by a factor of five and were averaged to characterize the PM emissions from drylots.

Relevant work was overlooked by EPA (2001a) for the estimation of cattle feed yard PM emissions. Recent work by Holmen et al. (2001) using lidar (light detection and ranging) was not included. The Parnell et al. (1994) study was not cited, but it is not clear whether that work was included in USDA (2000b), which was cited. Potential PM emissions from land spraying with treatment lagoon effluent are assumed to be negligible and thus were not considered further by EPA (2001a).

For PM, unlike most other air pollutants, emission factors developed for use in emission inventories and for dispersion modeling can, ideally, be reconciled using receptor modeling techniques. Receptor modeling makes use of the fact that atmospheric PM is composed of many different chemical species and elements. The sources contributing to ambient PM in an airshed also have specific and unique chemical compositions. If there are several sources and if there is no chemical interaction between them that would cause an increase or decrease, then the total PM mass measured at a "receptor" location will be the sum of the contributions from the individual sources. By analyzing the PM for various chemical species and elements, it should then be possible to back-calculate the contributions from various sources in the airshed. A variety of techniques are available for doing this; some (e.g., the chemical mass balance model; Watson et al., 1997) rely on the availability of predetermined source chemical composition libraries and are based on regression to determine the amounts contributed by various sources. Other receptor models are based on multivariate techniques and do not require source "fingerprints" determined a priori, but do require large numbers of receptor samples so that statistical methods can be applied. Target transformation factor analysis (Pace, 1985) and positive matrix factorization (Ramadan et al., 2000) are two examples of multivariate techniques that do not require explicit source composition data. Source apportionment may be especially useful for understanding the contributions from AFOs to the ambient PM in an airshed. Both receptor and dispersion modeling are associated with a significant level of uncertainty. The best approach is to use a combination of methods and attempt to reconcile their results.

VOLATILE ORGANIC COMPOUNDS

Emissions of volatile organic compounds (VOCs) from stationary and biogenic sources are significant, but limited data are available in most regions of the world. This situation makes it difficult to determine the impact of VOCs on a global basis. However, the United States (EPA, 1995a) and Europe have accumulated extensive data on the quantities and sources of their VOCs emitted to the atmosphere.

The three references in EPA (2001a) on VOC emission factors—Alexander, 1977; Brock and Madigan, 1988; and Tate, 1995—came from microbiology textbooks. Thus, the basis for determining VOC emission factors was rather weak.

Despite the paucity of data, attempts are being made to shed light on the estimation of emission factors for VOCs. For example, some for pesticides have been determined by the Environmental Monitoring Branch of the Department of Pesticide Regulation in Sacramento, California (California Environmental Protection Agency, 1998, 1999, 2000). The applicability of these efforts to VOC emissions from AFOs is unknown at this time.

Ongoing studies to determine emission rates of VOCs were not included in EPA (2001a). Scientists from Ames, Iowa, have developed techniques to collect and measure VOCs emitted from lagoons and earthen storage systems (Zahn et al., 1997). They found that 27 VOCs were prevalent in most samples and could be classified as phenols, indoles, alkanes, amines, fatty acids, and sulfur-containing compounds. Emission rates for many of these were determined at several sites, and the data have been transferred to EPA and state air quality specialists.

According to EPA (2001a), estimation of VOC emissions from confinement facilities, manure storage facilities, and manure application sites is difficult because of the lack of a reasonable method for estimating CH_4 production. CH_4

does not provide an appropriate basis for predicting VOC volatilization potential in livestock management systems. Gas-transfer velocities for CH_4 and VOCs differ by several hundredfold (MacIntyre et al., 1995). In addition, surface exchange rates for some VOCs are influenced by solution-phase chemical factors that include ionization (pH), hydrogen bonding, and surface slicks (MacIntyre et al., 1995). Physical factors such as temperature, irradiance, and wind are also major factors in the emission rates of sparingly soluble VOCs from liquid or semisolid surfaces (MacIntyre et al., 1995; Zahn et al., 1997). The differences in wind and temperature exposures between outdoor and indoor manure management systems can account for between 51 and 93 percent of the observed differences in VOC emissions (MacIntyre et al., 1995). This analysis suggests that exposure factors can account for differences observed in VOC flux rates, VOC air concentrations, and odor intensities. Therefore, the equation used to model the emission factor for VOCs in EPA (2001a) cannot be extrapolated for the majority of livestock operations.

Receptor modeling techniques can provide information on air quality impacts due to VOC emissions from AFOs. For example, Watson et al. (2001) reviewed the application of chemical mass balance techniques for VOC source apportionment. Multivariate methods have also been applied to source apportionment of ambient VOCs (Henry et al., 1995). Receptor modeling techniques to apportion VOCs from AFOs may be limited because many of the expected compounds may be formed in the atmosphere, react there, or have similar emission profiles from many sources.

To understand the contribution of AFO VOCs to ozone formation and gain insight into effective control strategies, measurements of individual compounds are essential. This is a difficult task because of the large number of compounds involved. The most widely used analytical technique involves separation by gas chromatography (GC) followed by detection using a flame-ionization detector (FID) or mass spectrometer (MS). The latter is useful for identification of nonmethane hydrocarbons using cryofocusing. VOC detectors that can be used for real-time measurements of typical ambient air are commercially available. New portable devices that use surface acoustic wave technology have been developed for field measurements of VOCs. Their sensitivity is not adequate to measure the low levels that may be harmful to humans. Research to support the development of more sensitive devices is needed.

There is a lack of information on the acute and chronic toxicological effects of VOCs from agricultural operations on children and individuals with compromised health. Recent epidemiological studies (without environmental measurements of VOCs) have shown higher incidences of psychological dysfunction and health-related problems in individuals living near large-scale swine production facilities (Schiffman et al., 1995; Thu et al., 1997). Further studies are needed to better understand the risks associated with human exposure to VOCs from AFOs.

ODOR

In a recent review, Sweeten et al. (2001) define odor as the human olfactory response to many discrete odorous gases. Regarding the constituents of animal odors, Eaton (1996) listed 170 unique compounds in swine manure odor, while Schiffman et al. (2001) identified 331. Hutchinson et al. (1982) and Peters and Blackwood (1977) identified animal waste as a source of NH_3 and amines. Sulfides, volatile fatty acids, alcohols, aldehydes, mercaptans, esters, and carbonyls were identified as constituents of animal waste by the National Research Council (NRC, 1979), and by Miner (1975), Barth et al. (1984), and the American Society of Agricultural Engineers (1999). Peters and Blackwood (1977) list 31 odorants from beef cattle feedlots. Zahn et al. (2001) found that nine VOCs correlated with swine odor. The sources of odors include animal buildings, feedlots, manure handling, manure storage and treatment facilities, and land applications.

Sweeten et al. (2001) also outline various scientific and engineering issues related to odors, including odor sampling and measurement methods. Odors are characterized by intensity or strength, frequency, duration, offensiveness, and character or quality. Odor concentration is used for odor emission measurement. Several methods are available for measuring odor concentrations including sensory methods, measurement of concentrations of specific odorous gases (directly or indirectly), and electronic noses.

Human sensory methods are the most commonly used. They involve collecting and presenting odor samples (diluted or undiluted) to panelists under controlled conditions using scentometers (Huey et al., 1960; Miner and Stroh, 1976: Sweeten et al. 1977, 1983, 1991; Barnebey-Cheny, 1987), dynamic olfactometers, and absorption media (Miner and Licht, 1981;Williams and Schiffman, 1996; Schiffman and Williams, 1999). Among sensory methods the Dynamic Triangle Forced-Choice Olfactometer (Watts et al., 1991; Ogink et al., 1997; Hobbs et al., 1999) appears to be the instrument of choice. Currently, there is an effort among researchers from several universities, including Iowa State University, the University of Minnesota, Purdue University, and Texas A & M University, to standardize the measurement protocol for odor measurement using the olfactometer.

Some odor emission data are available in the literature, particularly for swine operations (e.g., Powers et al., 1999). However, there are discrepancies among the units used in different studies. Standard measurement protocols and consistent units for odor emission rates and factors have to be developed. As shown in a recent review (Sweeten et al., 2001), the data (see Table L-1) on odor or odorant emission rates, flux rates, and emission factors are lacking for most livestock species (and for different ages and housing) and are needed for the development of science-based abatement technologies. Further research in well-equipped laboratories is needed as a precursor to rational attempts to develop emission factors for odor and odorants.

TABLE L-1 Odor Emission Rates from Animal Housing as Reported in the Literature

Animal Type	Location	Odor Emission Flux Rate (OU/s-m^2)[a]	Reference
Nursery pigs (deep pit)	Indiana	1.8[a]	Lim et al., 2001
Nursery pigs[b]	Netherlands	6.7	Ogink et al., 1997; Verdoes and Ogink, 1997
Nursery pigs	Minnesota	7.3-47.7	Zhu et al., 1999
Finishing pigs	Minnesota	3.4-11.9	Zhu et al., 1999
Finishing pigs[c]	Netherlands	19.2	Ogink et al., 1997; Verdoes and Ogink, 1997
Finishing pigs[d]	Netherlands	13.7	Ogink et al., 1997; Verdoes and Ogink, 1997
Finishing pigs (daily flush)[e]	Indiana	2.1	Heber et al., 2001
Finishing pigs (pull-plug)[e]	Indiana	3.5	Heber et al., 2001
Finishing pigs (deep pit)	Illinois	5.0	Heber et al., 1998
Farrowing sows	Minnesota	3.2-7.9	Zhu et al., 1999
Farrowing sows	Netherlands	47.7	Ogink et al., 1997; Verdoes and Ogink, 1997
Gestating sows	Minnesota	4.8-21.3	Zhu et al., 1999
Gestating sows	Netherlands	14.8	Ogink et al., 1997; Verdoes and Ogink, 1997
Broilers	Australia	3.1-9.6	Jiang and Sands, 1998
Broilers	Minnesota	0.1-0.3	Zhu et al., 1999
Dairy cattle	Minnesota	0.3-1.8	Zhu et al., 1999

NOTE: Rates have been converted to units of OU/s-m^2 for comparison purposes, where OU = odor unit.

[a]Net odor emission rate (inlet concentration was subtracted from outlet concentration).

[b]Number of animals calculated from average animal space allowance.

[c]Pigs were fed acid salts.

[d]Multiphase feeding.

[e]Odor units normalized to European odor units based on *n*-butanol.

SOURCE: Adapted from Sweeten et al. (2001).

About the Authors

Perry R. Hagenstein, Ph.D. (Chair), is a consultant on resource economics and policy and president of the Institute for Forest Analysis, Planning, and Policy, a nonprofit research and education organization. Prior to this, he was executive director of the New England Natural Resources Center and served as a Charles Bullard Research Fellow at the John F. Kennedy School of Government at Harvard. He also served as senior policy analyst for the U.S. Public Land Law Review Commission and was a principal economist for the U.S. Department of Agriculture (USDA) Forest Service. Hagenstein received his B.S. (1952) from the University of Minnesota, M.F. (1953) from Yale University, and Ph.D. (1963) in forest and natural resources economics from the University of Michigan. He currently serves on the National Research Council Board on Agriculture and Natural Resources and previously served on the Board on Earth Sciences and Resources and Board on Mineral and Energy Resources. Hagenstein has served on ten prior National Research Council committees including the Committee on Noneconomic and Economic Value of Biodiversity: Application for Ecosystem Management, Committee on Hardrock Mining on Federal Lands (chair), Committee on Onshore Oil and Gas Leasing (chair), and Committee on Abandoned Mine Lands (chair).

Robert G. Flocchini, Ph.D. (Vice Chair), is professor of the Department of Land, Air and Water Resources and director of the Crocker Nuclear Laboratory at the University of California, Davis. His interests include the identification, transport, and fate of particulate matter with regard to agricultural sources and application of nuclear techniques for emission measurement and characterization in agriculture and environment. He received his B.A. (1969) from the University of San

Francisco and his M.A. (1971) and Ph.D. (1974) in physics from the University of California, Davis. Flocchini currently serves as a member of the USDA Task Force on Agricultural Air Quality and trustee of the National Institute for Global Environmental Change.

John C. Bailar III, M.D., Ph.D., is professor emeritus at the University of Chicago. He is a retired commissioned officer of the U.S. Public Health Service and worked at the National Cancer Institute for 22 years. He has also held academic appointments at Harvard University and McGill University. Dr. Bailar's research interests include assessing health risks from chemical hazards and air pollutants and interpreting statistical evidence in medicine, with a special emphasis on cancer. He received his B.A. (1953) from the University of Colorado, M.D. (1955) from Yale University, and Ph.D. (1971) in statistics from American University. He is a member of the Institute of Medicine and has served on more than 20 National Research Council committees including the Committee on Estimating the Health-Risk-Reduction Benefits of Proposed Air Regulations (chair), Committee on Risk Assessment of Hazardous Air Pollutants, and Committee on Epidemiology of Air Pollutants.

Candis Claiborn, Ph.D., is an associate professor in the Department of Civil and Environmental Engineering at Washington State University. Prior to that, she was a senior process control engineer at ARCO Petroleum Products and a process engineer at Chevron. Her areas of expertise include airborne particulate matter measurement, characterization, and emissions, and air pollution control. She received her B.S. (1980) in chemical engineering from the University of Idaho and Ph.D. (1991) from North Carolina State University. Dr. Claiborn was a member of the Western Governor's Association's Western Regional Air Partnership Expert Panel on Windblown and Mechanically Generated Fugitive Dust, and a contributing author for the U.S. Environmental Protection Agency's (EPA's) Air Quality Criteria Development for Particulate Matter.

Russell R. Dickerson, Ph.D., is a professor and chair (effective July 1, 2002) of the Department of Meteorology at the University of Maryland, College Park. Prior to this, he worked at the National Center for Atmospheric Research and at the Max Planck Institute for Chemistry in Mainz, Germany. He received his A.B. (1975) from the University of Chicago, M.S. (1978) from the University of Michigan, and Ph.D. (1980) in chemistry from the University of Michigan. His areas of expertise include atmospheric chemistry, air pollution, and biogeochemical cycles with an emphasis on NO_x, ozone, carbon monoxide, black carbon, and ammonia. Dickerson previously served on the National Research Council Panel to Review the Langley Distributed Active Archive Center (DAAC) and U.S. Mideast Research Grants Panel.

James N. Galloway, Ph.D., is a professor in the Department of Environmental Sciences at the University of Virginia and is currently a visiting scientist at the Marine Biological Laboratory and the Woods Hole Oceanographic Institution. His major interests include the biogeochemistry of emissions, transport, and fate of nitrogen and sulfur and their potential effects on ecology. He received his B.A. (1966) from Whittier College and Ph.D. (1972) in chemistry from the University of California, San Diego. Galloway has given expert testimony to state and federal agencies and legislatures on environmental issues. He has previously served on the National Research Council Global Climate Change Study Panel (Chair), Panel on Processes of Lake Acidification, Tri-Academy Committee on Acid Deposition, and Committee on Transport and Transformation Chemistry in Acid Deposition.

Margaret Rosso Grossman, Ph.D., J.D., is a professor of agricultural law in the Department of Agricultural and Consumer Economics at the University of Illinois. She has spent sabbatical leaves (1986-1987, 1993-1994, 2000-2001) and many summers in the Law and Governance Group (formerly Department of Agrarian Law) at Wageningen University, Netherlands. Her research interests include domestic and international agricultural and environmental law. She received her B. Mus. (1969) from the University of Illinois, A.M. (1970) from Stanford University, Ph.D. (1977) from the University of Illinois, and J.D. (1979) from the University of Illinois. Grossman is past president (1991) of the American Agricultural Law Association (AALA) and received the AALA Distinguished Service Award (1993). She was awarded the Silver Medal of the European Council for Agricultural Law (1999), and she has received three Fulbright grants to support her research in Europe. Grossman is a member of the bar in Illinois and the District of Columbia (inactive).

Prasad Kasibhatla, Ph.D., is an associate professor in the Division of Environmental Science and Policy at Duke University. His areas of expertise include tropospheric chemistry and transport, global tropospheric oxidants, global tropospheric aerosols, regional air quality, anthropogenic impacts on atmospheric composition and ecosystems, and global and regional tropospheric chemistry modeling. He received his B.S. (1982) from the University of Bombay, M.S. (1984) from the University of Kentucky, and Ph.D. (1988) in chemical engineering from the University of Kentucky. Dr. Kasibhatla has previously served on the National Aeronautics and Space Administration (NASA) Committee for Measurement of Air Pollution from Satellites and proposal review panels for National Oceanic and Atmospheric Administration (NOAA) and Department of Energy (DOE) atmospheric chemistry programs.

Richard A. Kohn, Ph.D., is an associate professor in the Department of Animal and Avian Sciences at the University of Maryland. His areas of expertise include the environmental impact of animal production systems, the effect of diet on nitrogen and phosphorus excretion, and modeling of nutrient metabolism and whole-farm nutrient management. He received his B.S. (1985) from Cornell University, M.S. (1987) from the University of New Hampshire, and Ph.D. (1993) from Michigan State University, all in animal science. In 1999, Kohn gave an invited presentation on "Calculating the Environmental Impact of Animal Feeding and Management" to the National Research Council Committee on Animal Nutrition.

Michael P. Lacy, Ph.D., is a professor and chair in the Department of Poultry Science at the University of Georgia. His area of expertise is poultry, specifically, production and management, housing and equipment, ventilation, management in hot climates, and mechanical harvesting. Lacy received his B.S. (1974), M.S. (1982), and Ph.D. (1985) from the Virginia Polytechnic Institute and State University.

Calvin B. Parnell, Jr., Ph.D., P.E., is a Regents Professor of the Department of Biological and Agricultural Engineering at Texas A&M University. He has special expertise in the air pollution regulatory process, including permitting and enforcement of air pollution regulations. His research expertise includes pollutant measurements, dispersion modeling, emission factor development, and air pollution abatement. In addition, Dr. Parnell is known for his expertise in agricultural processing, grain dust explosions, and energy conversion of biomass. He received his B.S. (1964) from New Mexico State University, M.S. (1965) from Clemson University, and Ph.D. (1970) in environmental systems engineering from Clemson University. Parnell is a registered professional engineer in Texas, a fellow of the American Society of Agricultural Engineers, and a member of the Air and Waste Management Association. He has provided expert testimony to state and federal legislatures on agricultural air quality. Parnell has previously served on the Texas Air Control Board and currently serves on the USDA Task Force on Agricultural Air Quality. He currently receives research funding from a Texas Legislative Initiative on Air Pollution Regulatory Impacts on Agricultural Operations. Dr. Parnell teaches undergraduate and graduate courses in air pollution engineering.

Robbi Pritchard, Ph.D., is a professor in the Department of Animal and Range Sciences at South Dakota State University. His interests include beef feedlot management and ruminant nutrition. Pritchard received his A.A. (1975) from Black Hawk Junior College, B.S. (1977) and M.S. (1978) from Southern Illinois University, and Ph.D. (1983) in animal science from Washington State University. He previously served on Farmland Industries' University Advisory Board and was an ex officio member of the Board of Directors of the Dakota Feed Manufacturers.

Wayne P. Robarge, Ph.D., is a professor of soil physical chemistry in the Department of Soil Science at North Carolina State University. His research interests include studies of emissions of ammonia from swine lagoons, temporal and spatial patterns in ambient ammonia and ammonium aerosol concentrations, nitrogen budgets using Geographical Information Systems, and dry deposition of ammonia and ammonium aerosols to crop and woodland canopies. He received his B.S. (1969) and M.S. (1971) from Cornell University and his Ph.D. (1975) in soil science from the University of Wisconsin-Madison. He currently serves on the USDA Task Force on Agricultural Air Quality. He currently conducts research from the North Carolina State University Animal and Poultry Waste Management Center as part of "An Integrated Study of the Emissions of Ammonia, Odor and Odorants, Pathogens and Related Contaminants from Potential Environmentally Superior Technologies for Swine Facilities".

Daniel A. Wubah, Ph.D., is a professor of biology and associate dean of the College of Science and Mathematics at James Madison University. Prior to this, Wubah was chairperson of the Department of Biology at Towson University. His special expertise includes rumen microbiology and anaerobic zoosporic fungi. He received his B.S. and B.Ed. (1984) from the University of Cape Coast (Ghana), M.S. (1987) from the University of Akron, and Ph.D. from the University of Georgia (1990). Wubah previously served on the National Research Council Panel for Review of Proposals Under the AID (Agency for International Development) Research Grants Program for the Historically Black Colleges and Universities—Agriculture, Health, and Social Sciences. He is a member of the Board of Governors of the National Aquarium in Baltimore.

Kelly D. Zering, Ph.D., is an associate professor in the Department of Agricultural and Resource Economics at North Carolina State University. His special expertise is the economics of swine production and processing. He received his B.S. (1977) and M.S. (1980) from the University of Manitoba and his Ph.D. (1984) in agricultural economics from the University of California, Davis. Dr. Zering has extension responsibilities in the areas of swine management and marketing. He has completed research funded by EPA and the Animal and Poultry Waste Management Center, titled "Economic Analysis of Alternative Manure Management Systems." He currently conducts research on manure technology evaluation funded by the North Carolina Attorney General-Smithfield Agreement via the Animal and Poultry Waste Management Center.

Ruihong Zhang, Ph.D., is an associate professor in the Department of Biological and Agricultural Engineering at the University of California, Davis. Her main interests include control of gaseous and particulate emissions from animal feedlots, and wastewater treatment. She is a member of the USDA multistate research project NCR-189, "Air Quality Issues Associated with Livestock Facilities" and a member of the American Society of Agricultural Engineers Committee on Environmental Air Quality. Zhang received her B.S. (983) from Inner Mongolia Engineering University (China), M.S. (1986) from the Northeast Agricultural University (China), and Ph.D. (1992) from the University of Illinois at Urbana-Champaign. She has a U.S. patent approved (filed by the University of California, Davis) for a "Biogasification of Solid Wastes by Anaerobic phased Solids Digester System."

Board on Agriculture and Natural Resources Publications

POLICY AND RESOURCES

Agricultural Biotechnology and the Poor: Proceedings of an International Conference (2000)
Agricultural Biotechnology: Strategies for National Competitiveness (1987)
Agriculture and the Undergraduate: Proceedings (1992)
Agriculture's Role in K-12 Education (1998)
Agriculture's Role in K-12 Education: A Forum on the National Science Education Standards (1998)
Alternative Agriculture (1989)
Animal Biotechnology: Science-Based Concerns (2002)
Brucellosis in the Greater Yellowstone Area (1998)
Colleges of Agriculture at the Land Grant Universities: Public Service and Public Policy (1996)
Colleges of Agriculture at the Land Grant Universities: A Profile (1995)
Designing an Agricultural Genome Program (1998)
Designing Foods: Animal Product Options in the Marketplace (1988)
Ecological Monitoring of Genetically Modified Crops (2001)
Ecologically Based Pest Management: New Solutions for a New Century (1996)
Emerging Animal Diseases - Global Markets, Global Safety: A Workshop Summary (2002)
Ensuring Safe Food: From Production to Consumption (1998)
Exploring Horizons for Domestic Animal Genomics: Workshop Summary (2002)

Forested Landscapes in Perspective: Prospects and Opportunities for Sustainable Management of America's Nonfederal Forests (1997)
Frontiers in Agricultural Research: Food, Health, Environment, and Communities (2002)
Future Role of Pesticides for U.S. Agriculture (2000)
Genetic Engineering of Plants: Agricultural Research Opportunities and Policy Concerns (1984)
Genetically Modified Pest-Protected Plants: Science and Regulation (2000)
Incorporating Science, Economics, and Sociology in Developing Sanitary and Phytosanitary Standards in International Trade: Proceedings of a Conference (2000)
Investing in Research: A Proposal to Strengthen the Agricultural, Food, and Environmental System (1989)
Investing in the National Research Initiative: An Update of the Competitive Grants Program in the U.S. Department of Agriculture (1994)
Managing Global Genetic Resources: Agricultural Crop Issues and Policies (1993)
Managing Global Genetic Resources: Forest Trees (1991)
Managing Global Genetic Resources: Livestock (1993)
Managing Global Genetic Resources: The U.S. National Plant Germplasm System (1991)
National Research Initiative: A Vital Competitive Grants Program in Food, Fiber, and Natural Resources Research (2000)
New Directions for Biosciences Research in Agriculture: High-Reward Opportunities (1985)
Pesticide Resistance: Strategies and Tactics for Management (1986)
Pesticides and Groundwater Quality: Issues and Problems in Four States (1986)
Pesticides in the Diets of Infants and Children (1993)
Precision Agriculture in the 21st Century: Geospatial and Information Technologies in Crop Management (1997)
Professional Societies and Ecologically Based Pest Management (2000)
Rangeland Health: New Methods to Classify, Inventory, and Monitor Rangelands (1994)
Regulating Pesticides in Food: The Delaney Paradox (1987)
Resource Management (1991)
The Role of Chromium in Animal Nutrition (1997)
The Scientific Basis for Estimating Air Emissions from Animal Feeding Operations: Interim Report (2002)
Soil and Water Quality: An Agenda for Agriculture (1993)
Soil Conservation: Assessing the National Resources Inventory, Volume 1 (1986); Volume 2 (1986)
Standards in International Trade (2000)
Sustainable Agriculture and the Environment in the Humid Tropics (1993)

Sustainable Agriculture Research and Education in the Field: A Proceedings (1991)
Toward Sustainability: A Plan for Collaborative Research on Agriculture and Natural Resource Management
Understanding Agriculture: New Directions for Education (1988)
The Use of Drugs in Food Animals: Benefits and Risks (1999)
Water Transfers in the West: Efficiency, Equity, and the Environment (1992)
Wood in Our Future: The Role of Life Cycle Analysis (1997)

NUTRIENT REQUIREMENTS OF DOMESTIC ANIMALS SERIES AND RELATED TITLES

Building a North American Feed Information System (1995) (available from the Board on Agriculture)
Metabolic Modifiers: Effects on the Nutrient Requirements of Food-Producing Animals (1994)
Nutrient Requirements of Beef Cattle, Seventh Revised Edition, Update (2000)
Nutrient Requirements of Cats, Revised Edition (1986)
Nutrient Requirements of Dairy Cattle, Seventh Revised Edition (2001)
Nutrient Requirements of Dogs, Revised Edition (1985)
Nutrient Requirements of Fish (1993)
Nutrient Requirements of Horses, Fifth Revised Edition (1989)
Nutrient Requirements of Laboratory Animals, Fourth Revised Edition (1995)
Nutrient Requirements of Nonhuman Primates, Second Revised Edition (2003)
Nutrient Requirements of Poultry, Ninth Revised Edition (1994)
Nutrient Requirements of Sheep, Sixth Revised Edition (1985)
Nutrient Requirements of Swine, Tenth Revised Edition (1998)
Predicting Feed Intake of Food-Producing Animals (1986)
Role of Chromium in Animal Nutrition (1997)
Ruminant Nitrogen Uses (1985)
The Scientific Basis for Estimating Air Emissions from Animal Feedings Operations: Interim Reort (2002)
Scientific Advances in Animal Nutrition: Promise for the New Century (2001)
Vitamin Tolerance of Animals (1987)

Further information, additional titles (prior to 1984), and prices are available from the National Acadmies Press, 500 Fifth Street, NW, Washington, D.C. 20001, 202-334-3313 (information only). To order any of the titles you see above, visit the National Academy Press bookstore at *http://www.nap.edu/bookstore*.